Future of Business and Finance

The Future of Business and Finance book series features professional works aimed at defining, analyzing, and charting the future trends in these fields. The focus is mainly on strategic directions, technological advances, challenges and solutions which may affect the way we do business tomorrow, including the future of sustainability and governance practices. Mainly written by practitioners, consultants and academic thinkers, the books are intended to spark and inform further discussions and developments.

Abhishek Gupta ·
Dwijendra Nath Dwivedi · Jigar Shah

Artificial Intelligence Applications in Banking and Financial Services

Anti Money Laundering and Compliance

Abhishek Gupta
Effiya Technologies Private Limited
Singapore, Singapore

Dwijendra Nath Dwivedi
Department of Economics and Finance
Cracow University of Economics
Kraków, Poland

Jigar Shah
Effiya Technologies
Ahmedabad, India

ISSN 2662-2467 ISSN 2662-2475 (electronic)
Future of Business and Finance
ISBN 978-981-99-2570-4 ISBN 978-981-99-2571-1 (eBook)
https://doi.org/10.1007/978-981-99-2571-1

This Springer imprint is published by the registered company Springer Nature Singapore Pte Ltd.
The registered company address is: 152 Beach Road, #21-01/04 Gateway East, Singapore 189721, Singapore

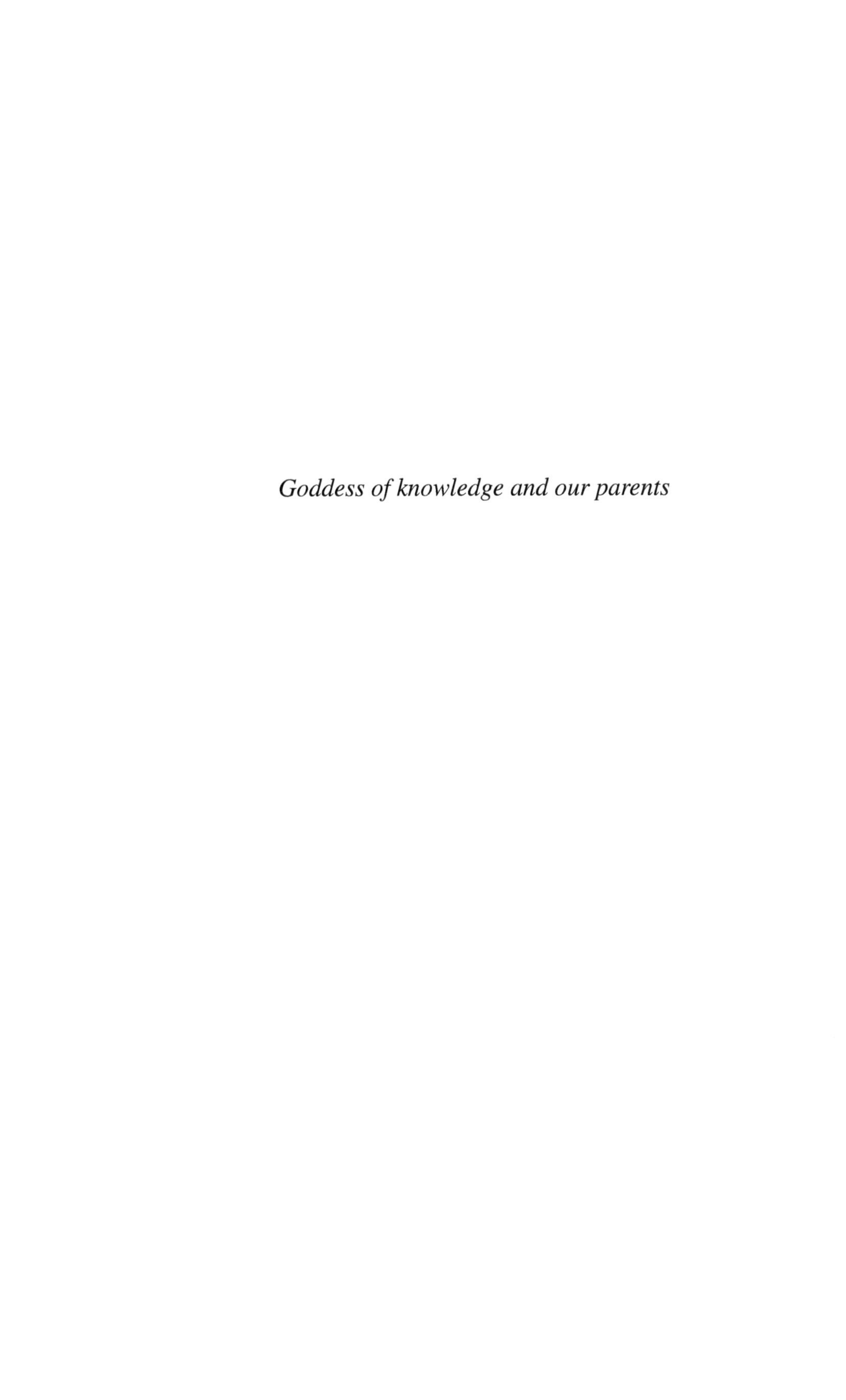

Goddess of knowledge and our parents

Acknowledgments

Writing a book on this topic was inconceivable for me till a few years ago. I have been a consultant throughout my career with a bit of flair for teaching and writing for journals. It was all about crisp communication, and condensing thoughts for senior management, who have an attention span of not more than five minutes. Transitioning from there to a detailed book has been a journey. Throughout the journey, I was inspired by my wife Liudmyla, who kept me motivated, made nice exhibits, and proofread for me. It was her support that made me finish the book on time. I would also like to acknowledge my mom and sisters who always motivated me to achieve higher goals and never settle for mediocrity in life. They have always been a source of inspiration for me. A special thanks to my team Ashish Jain, Indrani Biswas, Ravi Saroj, and Jay Modi, who provided all the necessary data-driven research, proof of concept, and other analyses, that helped me provide a few practical perspectives in this book. Last but not least, a special thanks to the Springer team that helped us throughout the process and kept providing guidance on topics that were new to us as authors.

I, along with my co-authors, would like to acknowledge the contribution of the following professionals, who have provided their valuable input and insights:

Mr. Jas Anand, Senior Partner Risk, Advisory, Ernst & Young LLP, Canada
Mr. Victor Matafonov, Head of Group Compliance, Emirates NBD, UAE
Mr. Abhishek Jhunjhunwala, Head of FCC Advisory, Tata Consultancy Services, India
Mr. Nipun Srivastava, Managing Director FCC advisory, Protiviti Middle East
Mr. Dawn Thomas, Director, Governance Risk and Compliance, Crowe UAE
Mr. Amit Keshri, Chief Compliance Officer, Bank of Baroda, UAE
Mr. Kantilal Bhati, Head of Compliance, National Bank of Oman, Oman

Abhishek Gupta
Singapore, Singapore

Contents

About the Authors

Abhishek Gupta possess over 18 years of experience in analytics driven advisory, with focus on enterprise-wide risk management, forensics for financial crimes and corporate strategy. Abhishek was also the risk management expert for McKinsey & Co. and then with Sutra Management Consultancies, where he has successfully worked with over 30 banks and financial institutions on Risk and Compliance offerings, South East Asia, North America and Europe. Abhishek has been working with his team on new emerging technologies like text analytics, voice and image analytics. Academically, he has also been one of the co-inventors of a provisional patent on fraud management technology in India, authored few research papers in reputed journals and has been a visiting faculty for MBA colleges.

Dwijendra Nath Dwivedi is a seasoned professional with over 19 years of expertise in creating valuable propositions for analytics and AI. He holds a post-graduate degree in Economics from Indira Gandhi Institute of Development and Research, and he is currently pursuing a PHD from Krakow University of Economics in Poland. He has presented his research at more than 20 international conferences and has published numerous Scopus indexed papers on the adoption of AI across various domains. As an author, he has made significant contributions to more than 8 books and has published over 25 impactful articles in renowned journals. Dwijendra conducts seminars and workshops on AI value for executive audiences and power users. In his capacity as a thought leader, he actively bridges the gap between business requirements and acts as a catalyst for enhancing analytical capabilities. His efforts cultivate a culture of analytical thinking, which in turn drives the development of effective business strategies.

Jigar Shah is a techno-management professional with 12 years of work experience into BFSI domain in business and analytics, consulting, IT services, project management and private equity. He carries hands-on experience in executing challenging assignments and consulting clients in areas of financial risk, compliance, and business intelligence. He has a rich experience in working with teams and clients across geographies.

Acronyms

AML	Anti-Money Laundering
API	Application Programming Interface to integrate solutions
ATL	Above the Line
BSA	Bank's Secrecy Act
BTL	Below the Line
CAGR	Compounded Annual Growth Rate
CCO	Chief Compliance Officer
CDD	Customer Due Diligence
CFT	Combating Financing of Terrorism
CRS	Common Reporting Standards
CTR	Currency Transaction Report filed by a financial institution to the central bank
e-KYC	Know your customer by leveraging digital tools
ETF	Electronically Traded Fund
EU	European Union
FATCA	Foreign Account Tax Compliance Act
FATF	Financial Action Task Force
FBI	Federal Bureau of Investigation
FCC	Financial Crime Control
FinCEN	Financial Crimes Enforcement Network
FIs	Financial Institution
FIU	Financial Investigation Unit
FTE	Fulltime equivalent—used for measuring productive effort of one human day
G-7	Group of seven countries
GDPR	General Data Protection Regulation
GINI	A coefficient ranging between 0 and 1 that provides indication of predictive power
HNI	High-net-worth individuals
IT	Information Technology
KS Statistic	A statistic ranging between 0 and 1 that provides indication of predictive power
KYC	Know Your Customer
MVP	Minimum Viable Product

NBFC	Non-Banking Finance Company
NLG	Natural Language Generation
NLP	Natural Language Processing
OCC	Office of the Comptroller of the Currency
OFAC	Office of foreign asset control
PEP	Politically exposed person
Regtech	Company specializing in technology for regulatory compliance
SAR	Suspicious activity report filed by a financial institution to the central bank
SME	Small and medium enterprises

Overview of Money Laundering

1

1.1 Introduction

Money laundering is an act of disguising illegal or tax-avoided money and bringing it into formal monetary channels. It primarily originates from two key sources. The first is money generated from various criminal activities like drug trafficking, human trafficking, bribery, or other illegal business. The second significant source of money laundering is the funds from tax avoidance. Governments in several countries impose a tax on income earned by individuals and non-individuals, such as corporations and partnerships. There exist beneficiaries of this wealth who do not want to pay their due taxes. They use multiple methods to avoid paying taxes and resort to money laundering.

Over the last few years, financial institutions are increasingly tracking and investigating another money stream, terrorist financing. Unlike money laundering, participants of this activity have legally earned money. This money is routed to terrorist organizations by sympathizers of the terrorists' ideology.

As discussed, crimes are not new in this world and so is the mechanism that people use to stash the money sourced by the wrongdoings. The unconfirmed reports suggest the existence of money laundering before the advent of the formal economy came into being. This was practiced by Chinese traders to avoid paying taxes. A more formal form of money laundering has been prevalent since the 1930s in the USA. A big chunk of authors has mentioned Al Capone, who intends to evade drug proceeds from the US authorities.

From the 1930s till the 1980s, there was a fair evolution of regulations, the focus of regulators on launderers. It got prominence in the 2000s and in the last few years, regulators are increasingly tightening their grip on not only the perpetrators of these crimes. The regulators are also very severe on the financial institutions that keep a blind eye on this category of customers and enable them while become a part of this dirty money.

A. Gupta et al., *Artificial Intelligence Applications in Banking and Financial Services*, Future of Business and Finance, https://doi.org/10.1007/978-981-99-2571-1_1

Another distinction for the purpose of the book would be to draw a line on the focus of this book. Financial crimes in general entail multiple types of crimes like fraud, money laundering, terrorist financing, cybercrimes, and so on. While compliance is now expanding its ambit to focus on all such crimes, the way artificial intelligence be applied to topics like fraud, money laundering, and cybercrime would be very different. For the book, the focus will primarily be on money laundering and terrorist financing. The way it is perpetrated and the ways through which organizations monitor, mitigate, and control these activities.

1.2 Overview of Various Type of Money Laundering

Money laundering is done for different purposes, depending on the source of money. There are two major types of money source, which requires laundering:

Illegal money—this is money sourced from illegal businesses. There are multiple illegal businesses like human trafficking, drug trafficking, bribery, ransom, money attained through transaction fraud, and many other similar types of activities.

These activities generate a substantial amount of unaccounted money that needs to be channelized through legal channels, for it to be consumed. The motivation of perpetrators of crime is to route this money through various channels and then deploy the money to assets or as savings in financial institutions.

Tax evasion, money used here is legal money, generated from genuine business activities. However, organizations intend to evade taxes. To avoid payment of taxes, the money is routed through complex structures, bogus routes are created, and then it is channelized back to the owner either in the host country or abroad.

The other source of money that is tracked and stopped from transiting is terrorist financing. Origin of terrorist financing is generally legally sourced money. It is donated or routed by various individuals and organizations who are sympathizers of the cause that those terrorist organizations stand for. Typically, these terrorist organizations have a façade organization or a face organization. The responsibility of such an organization is to engage in some humanitarian efforts. Money is genuinely routed to either such façade organizations or directly to the bank accounts of such terrorist organizations. Route followed is through countries that do not proscribe these establishments as terrorist organizations or there are cases where the control on the monitoring of money flow is weak. Money launderers try to leverage these weaknesses to eventually provide this money for certain terrorist activities, crimes against humanity, or a country.

Monetary transactions to watch listed or sanctioned entities—Lastly, governments and international organizations like United Nations, FBI, European Union, and similar law enforcement or regulatory organizations keep sanctioning individuals, organizations, and countries. It is illegal to deal with such sanctioned entities from the date of sanctioning. Financial institutions are mandated with the responsibility of ensuring that such entities are filtered when a customer account is created, or in case of money remitted, the counterparty should not belong to this sanctioned list.

These all form the core of AML and CFT monitoring.

1.3 Mechanism for Laundering Money

Money laundering organizations typically use multiple methods in their process. Some of the mechanisms are mentioned for a better understanding of the reader:

Cash transactions are used by most of human traffickers, drug traffickers, hawala transactors, small-scale tax evaders, and similar perpetrators. Modus operandi for them is simple. These businesses run on small-ticket cash on a regular basis. They keep bringing this money and depositing that to a consolidated group of beneficiaries from multiple locations.

Mule accounts. As the name suggests, this is an account used as a mule. It can be a new account opened with a forged identity or unsuspecting individuals who were rewarded to open a bank account in their names, or an existing customer of a financial institution, who is rewarded for transacting the money forward for a charge. In each of these transactions, the account is loaded with the inward remittance, and then it is either withdrawn in the form of cash or a big portion of the money is remitted outward. The remaining portion of the money is retained by the account holder as a reward for letting its account be used for muling.

Structuring. In structuring, the money is moved into the account or moved out of the account through various modes like cash, check, or wire transfers. However, they are moved in a coordinated manner in such a way that the transactions stay below the radar of "large ticket" transaction, where financial institution will seek clarification of the reason for conducting such a transaction.

Layering. Layering refers to transitioning money into different entities or in different investment instruments. For example, money collected from small tickets is converted into Electronic Traded Fund (ETF), casino chips, crypto assets, and so on. These are then converted into another investable asset. Once the money is converted into assets more than once, it becomes difficult to detect the original source of money. To make it more complex, money launderers also rotate the money globally by investing in these investable assets across geographies. Example of layering:

"Odebrecht is a Brazilian company that has executed multiple projects across the world. Their executives confessed that they paid bribes to receive preferential treatment for their bids. This money was paid through layering. Investigations revealed three layers in transactions (a) "off the books" money that is kept hiding in Swiss bank accounts (b) web of companies that are operating in multiple geographies and (c) inactive shell companies incorporated in Panama, which were used for paying the bribe". Source: Europol information.

Integration. Integration refers to transferring the money or asset in the money launderer account through legitimate money. After the layering, the money is eventually transferred to the money launderer. Simple due diligence of the money will

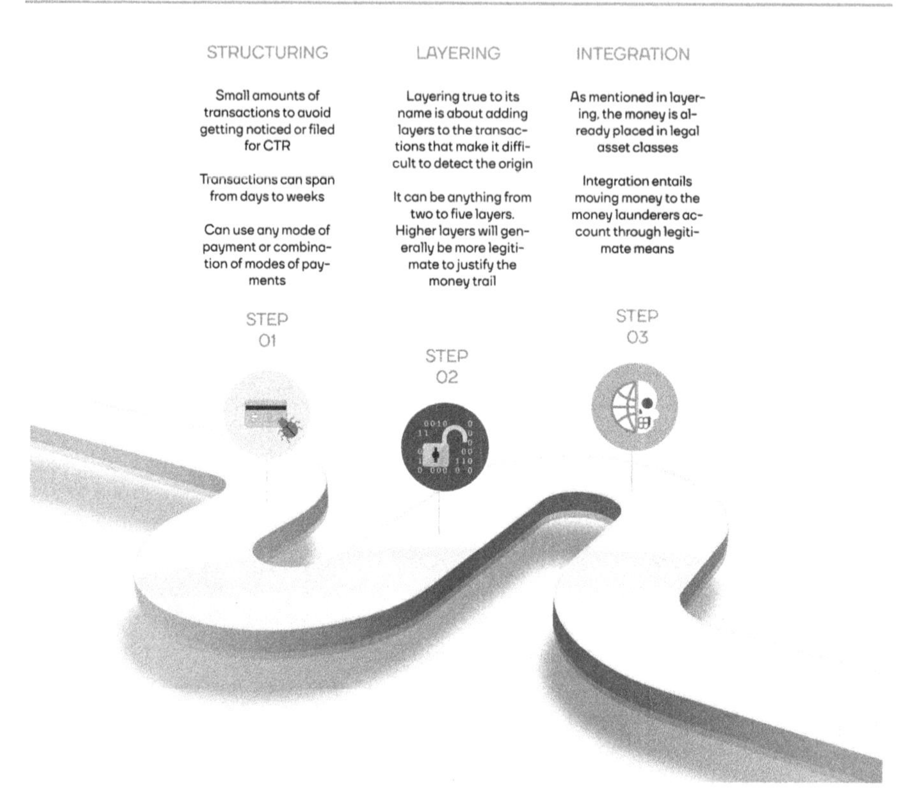

Fig. 1.1 Three steps of money laundering

often lead to legitimate means through which the asset is purchased, or the money is deposited (Fig. 1.1).

Purchase of financial instruments and foreclosures. This is another tool of money launderers. The money launderers purchase a financial asset, for example, a mortgage with a down payment in cash, an insurance policy with a premium payment in cash, purchase of a financial asset with a payment in cash. Once the asset is endorsed in the name of the person, the asset is liquidated prematurely. The proceeds of this premature liquidation are then forwarded to the individual through formal routes after deduction of foreclosure penalties.

Shell companies. This is a mechanism used by large-scale organizations either with the purpose of tax evasion or in case of large value transactions for bribery. In this case, there are:

- Legal companies that are created in the name of bogus identities
- Individuals whose identities are stolen
- Unsuspecting poor people who are asked to sign a few papers for a small financial reward. Then they get positioned as directors and shareholders

- Once these companies are created, the accounts are opened in the name of such entities. Then these companies' accounts are used for routing money, for example through bogus invoices. Money is routed through a web of shell companies to eventually be delivered to the ultimate beneficiary.

In one of the recently disclosed cases by European authorities, they mentioned how shell companies were used in Belgium. They were supported by accounts professionals and lawyers. Incorporated in high-risk sectors for money laundering like construction-related, cleaning services, hospitality industry, or companies involved in import or exports. These companies were generally operated by relatively young owners of foreign origin. They are generally new to the country. Legal professionals were supporting them in terms of pre-incorporation support, creation of business plans, managing profiles, and getting all necessary approvals.

In another case, even money laundering was supported by legal professionals in the Czech Republic.

Regulator's Response to Money Laundering

Combatting money laundering has not been a focus area for a non-negligible set of financial institutions till a few years back. The problem resides with the reality that the money launderers are also providing a big source of business and deposits or investments to these institutions. People in the organizations were sometimes blinded by the business value and tended to ignore the crime part of the equation. Let us share a few examples from our experiences.

Senior management of an exchange house walks into the compliance officer's room. Ask him to onboard a high-risk customer without doing enhanced due diligence, despite the customer ranked as high risk in their CDD scorecard.

Another example is a sandbox benchmarking exercise for a money exchange business that revealed a higher number of sanctions matches. The compliance team did not accept the findings. They said that the number of alerts due to matches has increased. Our team sat with them to manually go over ~ 100 cases that were incrementally generated. Their compliance team acknowledged that these are true positives and should ideally be investigated.

Basically, they had an existing solution that was giving them exact matches on their sanctioned list. They were missing a lot of matches from the watchlists. This is when their regulator is under pressure and is ramping up audits on the financial institutions. Their contention was that we have limited staff. If there are more alerts generated, their team does not have sufficient capacity to investigate. It could either be staff constraint or their fear of loss of business if they migrate to other solutions. However, they were ready to risk eventual detection and penalties by the regulator.

These are just a few examples of how businesses react to financial crimes. Anything that increases their costs or compromises with the revenues are not welcome change. They need enforcement.

Regulators across the globe sensed this and started pushing them. Now, there are country-level regulations alongside, and there is also an organization titled

"Financial Action Task Force (FATF)". It was set up as a part of G-7 countries summit with the purpose of examining and creating mechanisms for monitoring and combatting money laundering. After the September 11 attacks, the FATF mandate was expanded to also monitor and manage efforts to combat terrorist financing as well. Today, FATF oversees and is a flag bearer for money laundering, terrorist financing, and combating arms proliferation through the creation of regulatory and process-driven measures for the international financial system.

FATF in its current structure reviews the robustness of every country's ability to combat money laundering, counterterrorism efforts and make recommendations to ensure that their financial system and regulatory environment are more efficient. It primarily assesses the member countries on two broad parameters:

- Effectiveness—ability to have a robust framework that protects the financial system from financial crimes. It has 11 broad areas on which FATF assesses the effectiveness of the financial crime framework
- Technical compliance—this refers to a list of 40 compliance items that needs to be met. It contains assessment on law, regulation, and legal instruments that can control terrorism or its proliferation and money laundering.

It also prescribes a listing of countries, based on the perceived robustness or weakness in meeting such expectations.

1.3.1 Financial Crimes Combatting in the USA

The agencies managing the financial crimes in the USA are a combination of the US department of treasury's financial crimes enforcement network (FinCEN), the US department of treasury, office of foreign asset control (OFAC), and Office of the Comptroller of the Currency (OCC). These agencies coordinate for oversight of safe and sound financial transactions and asset ownerships.

In terms of regulations, the USA has a set of regulations for the countries across the world. The USA has adopted a set of regulations like Banks' Secrecy Act, US Patriot Act, Money Laundering Control Act, Foreign Account Tax Compliance Act (FATCA), Money Laundering Suppression Act, intelligence reform, terrorism prevention act, and suppression of the financing of terrorism convention implementation act, National Defense Authorization Act 2021. The ambit of coverage is all national banks, federal savings associations, federal branches, and agencies of foreign banks.

Bank's secrecy act provides for maintaining records and reporting for customers onboarded as well as reporting the customers doing more than USD 10,000 in aggregated cash transactions or purchasing negotiable instruments in cash. It was amended after the introduction of the US Patriot Act included maintaining the customer identification process. BSA seeks to manage—conducting customer due diligence, screen against a list of sanctioned or high-risk individuals prescribed by OFAC and other government lists, report suspicious activity, and manage the

AML program. The Money Laundering Control Act provides for making it a federal offense for any person or entity engaged in supporting transactions related to criminally derived money or assets. Money Laundering Suppression Act provides for expanding the ambit to reporting from depository institutions as well as including negotiable instruments issued by foreign countries.

Lastly, the suppression of the financing of terrorism convention implementation act is a part of the global convention. A section of the act makes any individual facilitating terrorist activities or possessing the knowledge of a fund being used for such activities an offender. Even acts that can result in the intimidation of the public for furthering the cause of such terrorist activities are also considered unlawful.

1.3.2 Financial Crimes Regulatory Evolution in the EU

For the European Union, this translated into a set of regulations in line with the recommendations from FATF. Over the last 30 years, the EU has come out with six directives and ensured that these directives are enforceable legally through a robust framework. Financial investigation units (FIU) are also defined along with guidelines on their collaboration across member states.

The first directive from the AML perspective was launched in 1991 after the constitution of FATF in 1989. As per the directive, it defined proceeds from criminal activity; before that, AML was always linked to drug cartels. It also required financial institutions in all EU countries to collect details for onboarded customers and track all transactions exceeding Euro 15,000. All suspicious activities are supposed to be reported to the relevant authorities.

The second directive focused on expanding the ambit of professions for money laundering as well as expanded coverage on the type of crimes that need to be tracked from a money laundering perspective. It included the responsibility of auditors, accountants, tax advisors, real estate agents, high-value items traders, and gambling as well as notaries and lawyers.

With the focus on terrorist financing and non-named accounts transactions, EU issued the third directive with a focus on banning no-name accounts and put the onus of conducting KYC and CDD on the FIs. There was also more stress put on terrorist financing.

The fourth directive issued in 2015 focused on further expanding the definition of financial crimes including tax crimes. The concept of high-risk countries was also introduced, and the EU pushed for regular assessment of AML and CFT efforts on both internal and cross-border transactions.

The fifth directive from the EU focused on providing more powers to the financial investigation units. It also focused on increasing transparency for ultimate beneficial ownership and trust. This was the approximate time when the virtual currency exchanges were launched. This directive included virtual currencies also in its ambit.

Sixth and the last updated directive from the EU has made a few additional changes (reference). First of them is the harmonization of predicament crimes (crimes that promote or are done to serve eventual money laundering crime). There are a total of 22 predicament crimes list prescribed. It includes crimes like insider trading, direct and indirect tax evasion. More importantly, there are additions like cybercrime and environmental crimes to make them more relevant to the changing environment. Another important development in the sixth directive is the expansion of the ambit of criminal activity. Earlier, only the ultimate beneficiary was the culprit. That has been modified to include people, abetting and facilitating the money laundering are also brought into the ambit. Even the punishment term is made stricter for the perpetrator of the criminal act.

The next change is the inclusion of legal persons, for example, organizations, into the ambit of a criminal act. It puts the onus on the management of the company to prove that someone from within was perpetrating without their knowledge. Their failure to stop the criminal act can result in even their suspension or closure of business in relevant jurisdictions.

While the EU has been tightening its financial crimes policy, there is also increasing cooperation and coordination among the member nation; however, a report from the European court of auditors (reference) suggests that the implementation of the AML/CFT framework varies across the member countries. Their coordination is also fragmented. Improvements and evolution are common in any regulatory framework, but it is important for the EU to make its frameworks less fragmented, more seamless, and commonly applicable across jurisdictions.

There are similar jurisdictions evolutions by all major regulators. Risk typologies, coverage of financial crimes, customer onboarding due diligence, customer dynamic risk assessment, risk score-based assessment, the inclusion of digital wallets and virtual currencies, prepaid cards, expansion of ambit to various players dealing in high-value assets, the inclusion of terrorism, the ambit of facilitation and abetment or knowledge of such acts, and tax evasion. These all have formed a part of the financial crimes' regulatory framework.

FATF also has been doing peer reviews on most of these dimensions for a listing of various countries.

The regulators sensing the agility of the money laundering and fraudsters to stay ahead of the regulations have started collaborating with the fintech and regtechs to enhance and sharpen their capability to detect and mitigate financial crimes. Most of the compliance professionals and consultants alike mentioned support from the regulator in various markets. And initiative varies from creating a country-level beneficiary owner base, access to cross institution transaction behavior, machine learning-driven financial intelligence, or network-driven identification of layering, regulators across the world are promoting these initiatives. They are also regularly speaking to regtech associations to keep themselves updated and posing challenge to the startups on their specific business problems that they face as regulators.

Regulators across the world are also becoming more participatory than earlier. They are regularly circulating more discussion documents before finalizing the policy making. It goes down to the level of use case definition. Hongkong Monetary Authority (HKMA), for example conducted a conference with the financial institutions participating. Outcome of conference was multiple use cases around financial crime that were brainstormed.

Another development is the regulators' increasing appetite of understanding the technology. Earlier they used to be technology agnostic. Now, there exists an active engagement and almost to the level of developing suptech (supervisor collaborated technology).

1.4 Financial Institution's Response to Combatting Money Laundering and Terrorist Financing

Financial institutions (FIs), especially banks, were mixed, depending on the stakeholder, one is dealing with. We had the privilege of speaking to all sides of the spectrum in the financial institutions as a consultant. There was a clear conflict of interest between business seekers and compliance professionals. During our interactions with business teams, their focus was on getting more business. One of the heads of specialized organizations in an ASEAN country clearly highlighted that their team had a target of mobilizing a billion-dollar in deposits in a year. Their contention was that while their team complied with the regulations to the best extent possible, issues like KYC and direct source of funds mobilized were done with an intent to ensure that the financial institution does not get exposed legally. They were not keen on doing stricter due diligence, as it might have meant a big business diverting itself to competition.

That changed in the last five years with a lot of global organizations getting penalties, far outstripping business benefits that this money could bring to the business. It also posed reputation risks for the financial institutions.

Today, organizations across the spectrum and across geographies have a clear focus on financial crime monitoring, mitigation, and control. Organizationally, there exists a board of directors overseeing the compliance function. Most of the heads of compliances have direct access to the board of directors. Relatively larger organizations have also been conducting regular sensitization and training for the board of directors and senior management on financial crimes.

Majority of the heads of compliance that we have been exposed to state that financial crimes used to be a line item in the risk and compliance audit review of the regulator. Now, there are dedicated audits conducted by the regulators. The intensity of audits has risen to the levels of risk audits. There are general audits and topical audits that assess preparedness for a specific type of financial crime risk.

There exists unanimity among the compliance professionals that we spoke to regarding the importance of the head of the compliance role. Their access to the board of directors, appreciation by senior management, and leadership on the importance of compliance role has resulted in the heads of compliance having higher expectations from the stakeholders.

1.5 The Outcome of Increasing Focus on Financial Crimes

With the evolving tightening of the regulator-pushed controls, financial institutions are now moving closer to monitoring and reporting suspicious behaviors. However, perpetrators of these crimes are ahead of monitoring authorities, both in terms of sophistication and the financial power that they possess to overcome the checks and balances put by the financial institutions (Fig. 1.2).

Irrespective, the catching game is also improving. Institutions are increasingly becoming alert and that is reflected in the growth rate of SAR filings by institutions in major locations that have witnessed an average increase of around 20% between the years 2019 and 2020. The USA has in fact seen an increase of over 40% in SAR filings between years 2020 and 2021. A big portion of these is coming from the non-banking organizations.

It is going to be a fight for smarter solutions to detect and catch these perpetrators in the long run. Throughout this book, we will discuss how artificial

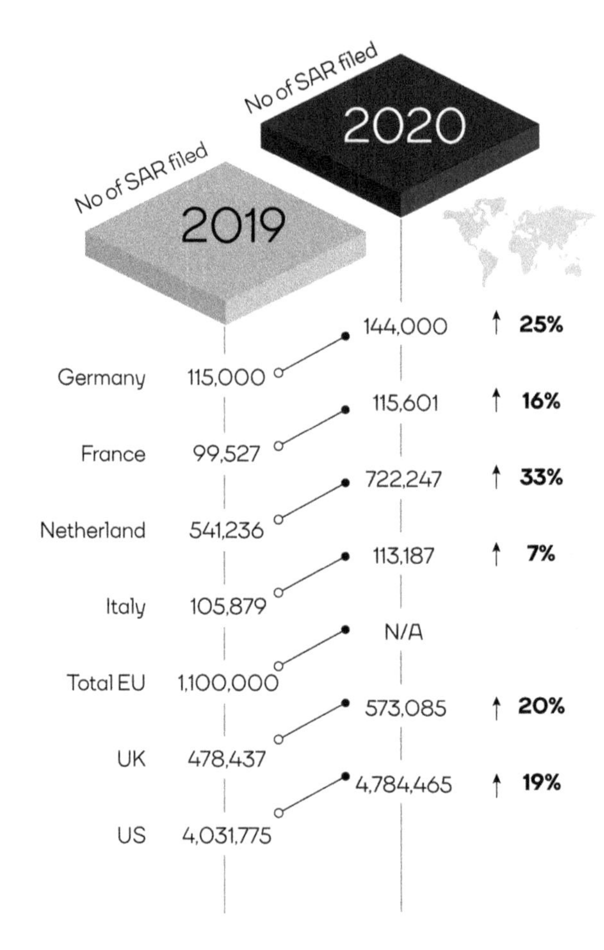

Fig. 1.2 Growth in number of SAR files in major economies

intelligence can play a role in detection and reporting while reducing manual efforts. The reader will go step by step on every topic of AML and terrorist financing management and by the end of the book will have a good understanding of business applicability and the key building blocks of leveraging AI.

References

https://www.eca.europa.eu/Lists/ECADocuments/SR21_13/SR_AML_EN.pdf
https://eur-lex.europa.eu/legal-content/EN/TXT/?uri=CELEX%3A52021PC0423
https://www.occ.treas.gov/topics/supervision-and-examination/bsa/index-bsa.html
https://www.congress.gov/bill/103rd-congress/house-bill/3235#:~:text=Money%20Laundering%20Suppression%20Act%20of%201994%20%2D%20Amends%20Federal%20law%20to,reporting%20requirements%20for%20depository%20institutions.
https://www.congress.gov/bill/99th-congress/house-bill/5077#:~:text=Money%20Laundering%20Control%20Act%20of%201986%20%2D%20Amends%20the%20Federal%20criminal,laundering%20as%20a%20Federal%20offense.
https://www.congress.gov/bill/107th-congress/house-bill/3275
https://www.govinfo.gov/content/pkg/PLAW-116publ283/pdf/PLAW-116publ283.pdf

Financial Crimes Management and Control in Financial Institutions

2

2.1 Introduction

Overseeing and managing anti-money laundering is highly regulated by the central banks. Central banks control this by providing a set of circulars related to transaction monitoring and sanctions monitoring. Central banks also conduct regular audits on bank preparedness and the intensity of these monitoring activities.

Any misses are severely dealt with in the form of audit comments, recommendations on tightening oversight through new scenarios, implementation of risk scores, and in select cases, promoting the application of machine learning-driven oversight. Central banks are also quite strict about time-bound plugging gaps in the audit observations.

Compliance organization is typically faced with multiple functions which are handled through regular monitoring investigations and reporting to the regulators.

Typical management of such things is done through setting up:

(a) Governance structure
(b) Active monitoring of financial crime events.

2.1.1 Governance Structure

A financial institution is expected to create a governance structure that has board-level oversight. The board of directors (BOD) is directly responsible for reviewing, approving, and implementing the financial crimes governance framework that comprises well-documented policies and procedures as well as the strategy. BOD should also institute a vigilance role and review any outcome of those for customers, employees, and business partners. This activity shall be conducted along with the efficacy of the control procedures and the breach incidences.

A. Gupta et al., *Artificial Intelligence Applications in Banking and Financial Services*, Future of Business and Finance, https://doi.org/10.1007/978-981-99-2571-1_2

BOD also appoints the head of compliance responsible for designing, implementing, and monitoring the financial crimes management structure.

Once the management structure is defined, it is the responsibility of the executives to create relevant policies and procedures for the detection, mitigation, and control of possible events that could be happening in the financial institution. For a larger organization, there can be a dedicated role of compliance officer that specifically focuses on the organization's oversight on monitoring of financial crimes management activities as well as compliance with regulatory guidelines in this regard.

The policies and procedures will essentially cover the following elements:

Policies

The policy should be based on the risk assessment of the business that looks at the client portfolio and the nature of the business relationships between the client and the institution. There are multiple elements to conduct the risk assessment: the products offered, the delivery channels, geographic locations of the clients and the transactions, technology platforms through which these products and services are offered or processed, and the nature of transactions. The policy will also assess and provide guidance for the activities of the partners and other intermediaries.

Policy needs to be well-documented, regularly updated based on risk assessment reviews or regulatory changes and approved by relevant authorities that are provided to all the employees. Employees dealing with monetary transactions, customers, agencies, etc. are also trained regularly to ensure that they understand and can effectively implement all the expectations set forth in the policy.

The policy will also clearly lay out the roles and responsibilities of the organization set up under the compliance officer. Specific elements for a robust policy will include the following but are not limited to:

(i) Requirements related to knowing your customer along with a well-documented approach to risk assessment of customer or key people of non-individual customers. If required, then information on related parties also will be assessed
(ii) Criteria for defining a high-risk customer and actions for conducting due diligence, follow-up, and monitoring requirements for such customers
(iii) Monitoring requirements from financial and non-financial transactions emanating from customers with the financial institutions
(iv) Requirements for capturing and storing critical information from all related parties and the transactions
(v) Reporting requirements to senior management and the regulator
(vi) Situations when a transaction must be stopped or blocked for fund transfer, virtual currencies-related transactions
(vii) Measures to assess the validity or reasonableness of a transaction like a customer supporting explanations, supporting documents, press searches, or other industry portal information solicitation, etc.

Procedures
The procedures will lay down detailed guidelines, forms, and requirements for uploading supporting documents for the following:

(i) Mechanism of customer risk assessment
(ii) Mechanism for investigation of a suspicious transaction including the query to be asked from a client, conducting open-source searches, looking at already available information including the one available in physical form and other types of secondary information coming from different agencies
(iii) Process of review of ongoing client risk assessment including the query to be asked from a client, conducting open-source searches, looking at information already present, including the one available in physical form and other types of secondary information coming from different agencies
(iv) Process and mechanism of reporting suspicious activity to senior management and to the regulator
(v) Process and mechanism for recommending discontinuation of relationship with a high-risk customer
(vi) KPI, frequency, and stakeholders for regular reporting.

Internal audit acts as the third line of defense to identify and report a possible breach of process or controls.

Another important element for the ongoing applicability and relevance of the framework of financial crime is the ongoing review. Depending on the country, regulators insist on the regular review of the anti-financial crime efforts, effectiveness in terms of the policies, process, training efficacy, and overall risk assessment to see if there are shifts and if there are needs for any adjustments in the program.

The weaknesses identified should be bridged from a governance point of view.

2.1.2 Active Monitoring of Financial Crime Events

Financial institutions are expected to monitor and manage the whole customer lifecycle to detect, mitigate, control, and report possible financial crime incidents.

One of the key domains handled within the compliance function is transaction monitoring—a real-time transaction screening for parties involved in the transactions. For money exchange houses, they need to observe a few of the rules in real-time before a remittance request is initiated. For many other institutions, it is restricted to checking the counterparty names and matching them against the watchlist, along with sanctions screening to ensure that the money transferred is not heading to the wrong hands.

A more comprehensive transaction monitoring, a check is typically executed at the end of the day. This monitoring considers various aspects of customer, payments, mode of payment, countries, location, correspondent bank, usage of products, and more monitoring. Various checks are done to ensure that the payments are not linked in any way to the proceeds of the financial crime.

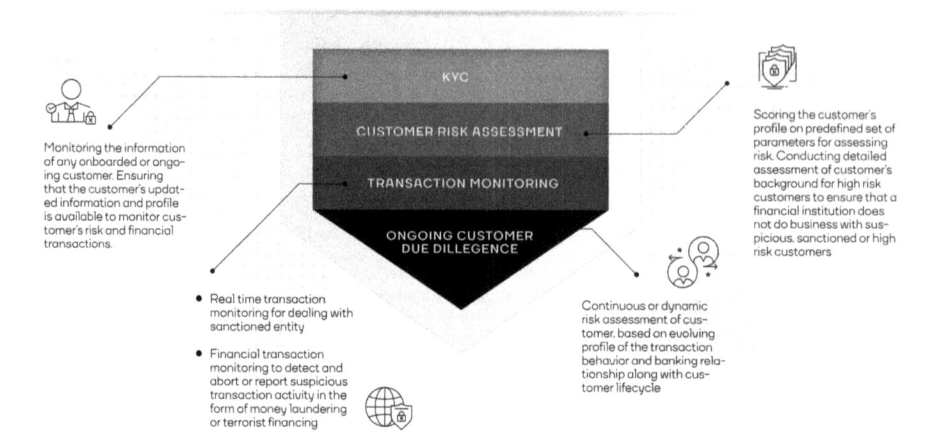

Fig. 2.1 AML CFT touchpoints during a customer lifecycle

Another activity that a compliance team takes care of is customer onboarding-related compliance. It includes onboarding customer risk due diligence, CRS, and FATCA compliance (Fig. 2.1).

Consumer protection, fraud-related transaction investigations are a few other activities that are handled by the compliance department. All these compliances which were less stringent in the past have already become a lot more challenging for the organizations due to regular and detailed reviews and penalties by the central banks. We would explain the steps and mechanism along with AI application in subsequent chapters. In this chapter, we would only introduce the core functions of financial crime management by the financial institutions throughout the lifecycle.

As shown in the exhibit 2.1, it starts during the customer onboarding. A match against a sanctioned list and application of a customer risk scorecard are the ways for the financial institution to ensure that the onboarded customer is of acceptable risk level from a financial crime's perspective.

The customer is also registered for FATCA and CRS reporting. FATCA is applicable for US tax residency determination and CRS is applicable internationally among countries to share account information between countries for combating tax evasion.

Once the customers are onboarded, their transactions start getting screened through real-time transaction monitoring (to assess if the ultimate beneficiary happens to be a sanctioned entity). There is also a set of rules administered in the form of scenarios that look at the various aspects of transaction behavior and flag anything suspicious.

Examples of a few of those transaction behaviors are listed in Figs. 2.2 and 2.3.

Organizations also undertake regular risk ratings of the customer based on the evolving banking relationship and the transaction behavior. This is called dynamic risk rating.

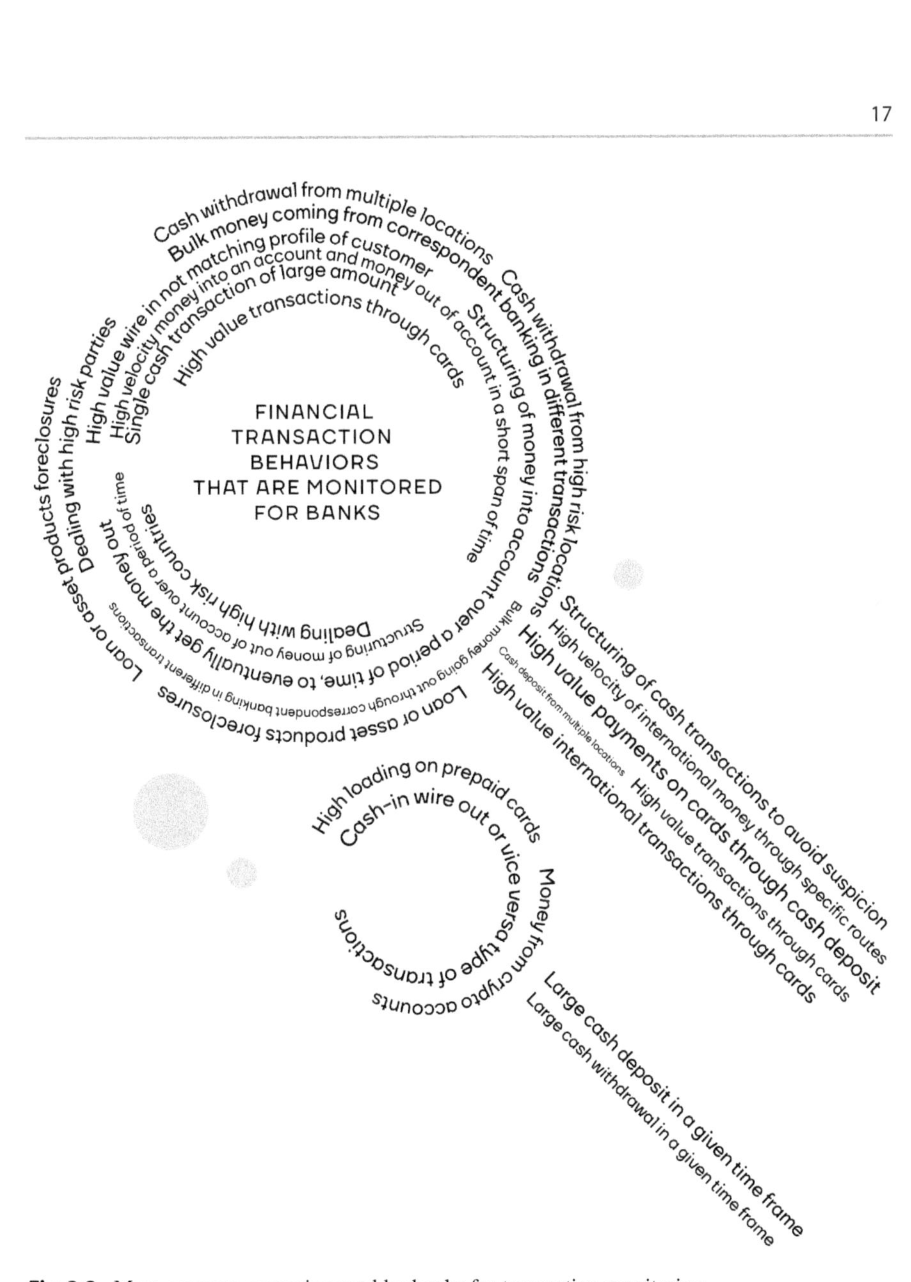

Fig. 2.2 Most common scenarios used by banks for transaction monitoring

Throughout the lifecycle, the onus of detecting, investigating, and reporting to regulators lies with the financial institutions. Regulators conduct regular checks on the robustness of the monitoring systems of the financial institutions and recommend time-bound fixes to the gaps identified.

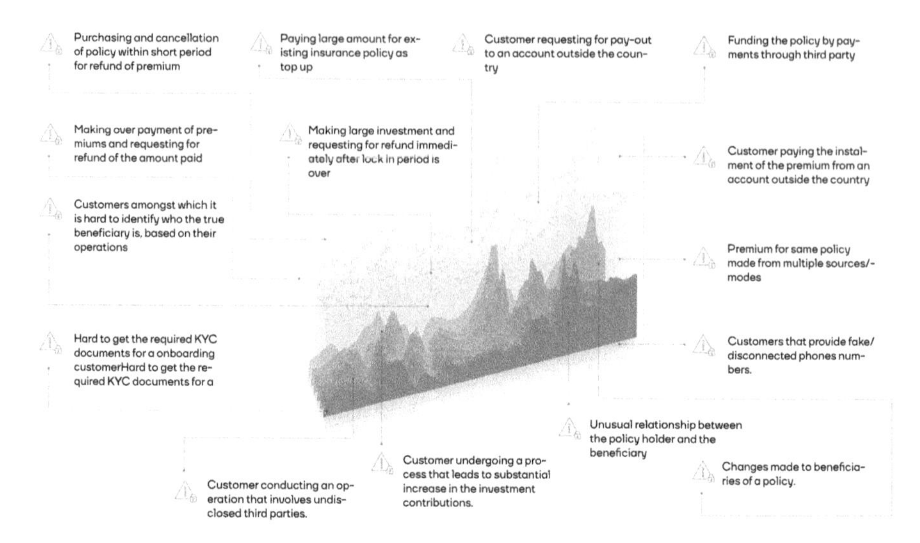

Fig. 2.3 Example scenarios for an insurance company

2.2 Organization Design for Financial Crimes

As explained above, the compliance department has more functions to cater to. However, we will focus primarily on the financial crime management organization. There are five broad functions that are covered within it.

2.2.1 Customer Risk Assessment

The customer risk assessment team primarily focuses on assessing customer risk rating from an onboarding perspective. The team is responsible for analyzing the customer details and providing an initial due diligence score. If the score is mapped to high risk, then conducting enhanced due diligence on the customer. As a part of enhanced due diligence, the team collects more information on the individual, related parties, and expected financial transactions or the historical financial transactions. The eventual outcome of the exercise is to ascertain whether the customer is of an acceptable risk level from the perspective of a financial crime. In case the customer risk is unacceptable—an extremely high risk of money laundering from the past incidences, then the financial institution will not enter into the relationship with the customer.

2.2.2 Sanctions Monitoring

This team investigates the cases related to watchlists, politically exposed persons (PEP), or blacklisted entities. It can be triggered during onboarding, or it could be a match on the counterparty for real-time financial monitoring. The objective of the team is to research the match and ascertain if the party is indeed a sanctioned entity. If the name matches a sanctioned entity or a there is an extremely high risk of money laundering, then such customers are reported to the central bank.

2.2.3 Financial Transaction Monitoring from an AML Perspective

Financial transactions monitoring team probes suspicious transactions done by various parties through different modes, patterns, countries, and channels. Any suspicious transaction exceeding a particular amount translates into an alert for the financial institution. Based on the prima-facie observation, the alert needs to be investigated.

2.2.4 Ongoing Customer Risk Assessment

Depending on the organization, this section of the department is merged with the customer risk assessment. For a few organizations, this is a separate team that caters to operational aspects of conducting Re-KYC and then rerunning the customer risk assessment framework.

2.2.5 Regulatory Reporting

This department focuses on liaison with the central bank. All audits, queries, and responses on query or audits are handled by this team. This team also focuses on the regular reporting of relevant information to the central bank.

2.2.6 Legal Team

Legal department's responsibility is to initiate legal actions emanating from all the above departments, pertaining to financial crime activity or the mitigation activity that the financial institution needs to initiate. The team also takes care of privacy laws and applicable customer secrecy while leveraging such data for business purposes.

2.2.7 Cybersecurity Team

This is a new department that is recommended within the compliance department. While it has a role to play in the investigation of the AML team, a fully dedicated team business case is more from the point of fraud management which also resides within the compliance. It implies that cybersecurity can be shared resourcing for compliance. A FINCEN official mentioned reports of over USD 5 billion in bitcoin payments that were associated with ransomware payments in 2021. There was a rising trend of such ransomware transactions paid over the last year.

The monitoring is centralized under the compliance function. However, a few of the activities like customer due diligence, updating KYC, and checking the reasons for suspicious transactions are also done by the branches. Organizations can also outsource some of the activities like conducting preliminary investigations in alerted cases, marking of suspicious activities to the frontline business teams for example branch relationship manager (RM) or customer service representative (CSR). It is more efficiently done by the frontline staffs they are in regular touch with the customers. They can do some of these activities and capture the required information to update the documents. The implication of this outsourcing would be that for a few cases on high-risk counterparties or the due diligence on the high-risk customers can be done by the business and then can be submitted to the centralized compliance for verification and taking relevant actions.

Another important aspect of organizing these functions is the scale and jurisdiction in which these entities operate. If we look at a relatively smaller organization like an exchange house with few branches, management of this function is typically centralized. For a larger organization, these activities follow a hybrid model—partly centralized and partly managed directly from the branches. The activities handled by the branches are typically managed by the business team.

In the case of international organizations that we call multinational financial institutions (MFI), this becomes more complicated. A few organizations have a hub and spoke model where the centrally organized entity is in the head office and each office has its own compliance team with the dotted line reporting to the central office or head office, whereas in other organizations, the units are quite independent depending on the scale and scope of the operations in that country. A relatively smaller branch office or a representative office will still be handled or managed from the head office. And the larger self-sustained entities will have a lot stronger and more independent compliance functions.

Even in terms of policies and processes, the trend is increasingly inclined toward centralization of the policies and processes. Earlier, organization units in different countries used to have a different set of policies and procedures for each office. As time evolved, the organizations realized that managing organizations of that scale in a completely scattered setup is going to be very difficult both in terms of oversight and control. Hence, the organizations (especially the multinational ones) are now a lot more aligned to a central mode of governance. These multinational organizations roll out centralized policies and procedures which are to be followed by businesses across countries. Alongside IT department, they also ask the local officers to comply with the local regulations but the highest or the stricter form of both. For example, if the central policy is looking at handling customer information, it will be strict enough for other geographies. However, in Europe, if GDPR compliance is to be followed, then for relevant countries' offices, GDPR compliances are implemented. For the remaining part of the organization, i.e., offices in different countries, it will still stay with stricter privacy customer privacy information but not necessarily GDPR.

Even in our view, the best possible structure for an organization will always be a more centrally controlled uniform set of policies and processes not only in terms of governance of the organization but also in terms of governance of data and of the technology and the best practices which must be percolated down even in terms of artificial intelligence.

In the last chapter titled "best-in-class AI-driven financial crimes management organization", we would recommend a desired organizational chart that will be of relevance. We understand that organizational design will always have a context. Having said that, we believe that there can be certain best-practice organizational structures that can work in an institution of any size.

2.3 Reporting Structure in Financial Organization

We have seen enough permutation and combinations for a compliance function governance in the older days. Compliance function was either handled by CFO or head of legal or in some of the cases chief risk officer. With time and with increased focus from the regulators (including evolution in policies and procedures like internal audit regulations), the compliance function has evolved even in terms of being a full-fledged entity.

Now, the organization is headed by a chief compliance officer who directly reports to the board of directors rather than even the CEO and the power and scope of a head of compliance are also increased significantly. Our recommended organizational structure will always have a head of compliance who should have direct access to the board of directors.

We conducted qualitative research on trends related to expectations of regulators and the boards of directors on the head of compliance officers. Compliance heads unanimously mentioned increased oversight from the regulators. The form of intensity varies by the financial institution, jurisdiction, and business; however, increased oversight remained a common theme. As per various compliance heads, financial crimes and compliance used to be a small subset of supervisory review, a few years back. Regulators used to view compliance as an allied function. Now, a review is dedicated to compliance. This has also changed the attitude of the board of directors on compliance as a function. The chief compliance officer is entrusted and is a part of the regular strategy and risk reviews of the board of directors.

This has had a direct impact on the roles, responsibilities, and size of the compliance organization.

Figure 2.4 is a summary of organizational structure or the roles explained by compliance heads for a mid to large-size organization. It has four broad functions. Each function is led by its head directly reporting to the chief compliance officer. They directly map to the roles that were explained for combatting money laundering and CFT by the financial institutions. The only difference is that customer risk assessment—be it at the time of onboarding and ongoing due diligence can be merged into a single department.

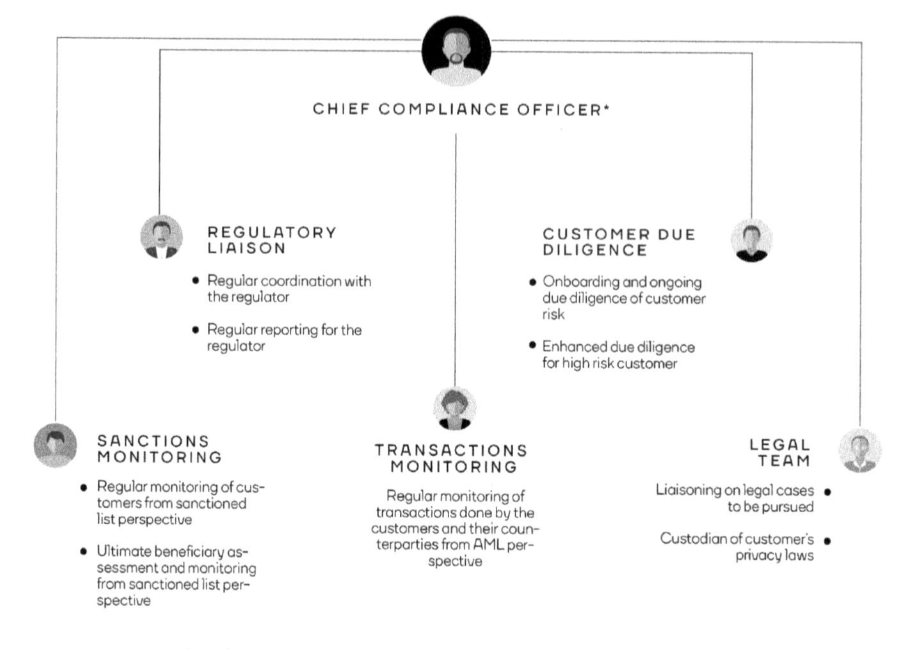

Fig. 2.4 Typical AML departments

As mentioned in the chapter above, cybersecurity is an emerging area which is housed in the compliance function now. Strategy or analytics functions are another emerging function. Strategy department leverages data and identifies the effectiveness of the whole organization from multiple perspectives: sharpness in rule-based solution, success of artificial intelligence-based monitoring, identifying ways of improving the function, etc. It also provides necessary business intelligence to the relevant stakeholders around emerging themes and trends across markets, and across businesses, so that FIs are better prepared to investigate and control the financial crimes. Lastly, it also provides intelligence on performance management in terms of various matrices like the responsiveness of the investigation team in closing cases, load on teams, aging cases.

2.4 Performance Management for the Compliance Team

Performance management for the compliance department is a summation of all the functions that form the department. Performance management for such a function is handled through a set of indicators that are mentioned in Fig. 2.5.

As mentioned above, the performance management for the financial crimes organization is broadly classified into following domains.

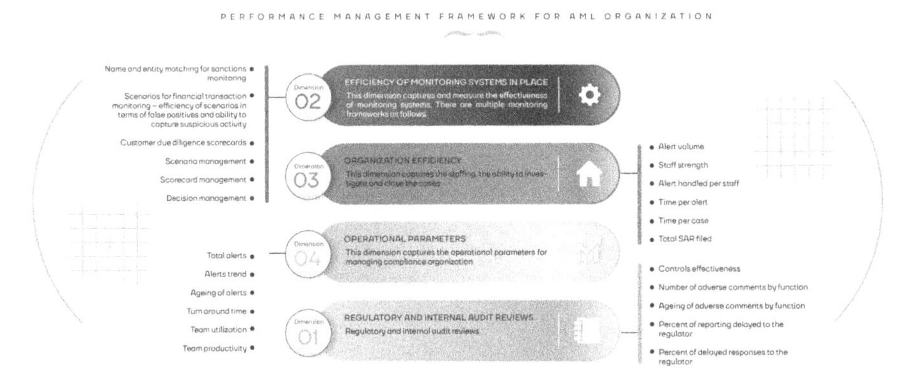

Fig. 2.5 Major KPIs for the AML department

2.4.1 Efficiency of Monitoring Systems in Place

This dimension captures and measures the effectiveness of monitoring systems. There are multiple monitoring frameworks as follows:

- Name and entity matching for sanctions monitoring
- Scenarios for financial transaction monitoring—efficiency of scenarios in terms of false positives and ability to capture suspicious activity
- Customer due diligence scorecards
- Scenario management
- Scorecard management
- Decision management.

2.4.2 Organization Efficiency

This dimension captures the staffing, the ability to investigate and close the cases:

- Alert volume
- Staff strength
- Alert handled per staff
- Time per alert
- Time per case
- Total SAR filed.

2.4.3 Operational Parameters

This dimension captures the operational parameters for managing compliance organization:

- Total alerts
- Alerts trend
- Aging of alerts
- Turnaround time
- Team utilization
- Team productivity.

2.4.4 Regulatory and Internal Audit Reviews

- Controls effectiveness
- Number of adverse comments by function
- Aging of adverse comments by function
- Percent of reporting delayed to the regulator
- Percent of delayed responses to the regulator.

For an organization, it is imperative that these indicators are not assessed on a standalone basis. They need to be assessed on self-benchmarking (evolution of the organization over a period) and industry benchmarking (industry parameters captured and then assessed for the goodness of performance).

With the advent of machine learning and artificial intelligence-driven functions, there is an additional set of reports that focus on the following are to be added:

- Model performance monitoring reports
- Model repository
- Model risk reporting.

Depending on the data organization set up for the FI, there could also be a few reports related to data governance that could be added.

With a sound organizational design and a robust performance management, we expect the financial crimes organization to deliver the desired value and ensure that the productivity and efficiency of the organization are maximized.

3 Overview of Technology Solutions

3.1 Introduction

A financial institution (FI) can unwittingly become a party to a financial crime during any stage of the customer lifecycle. Enterprise AML solutions focuses on every possible point of exposure. Owing to the sheer size and growth in the volume of global financial transactions, and the rare cases of FIs being complicit in instances of financial crime, regulators are becoming more proactive about the lack of robust checks in the sector.

FIs have responded by investing in financial crime detection strategies and investigation solutions. The solution space has witnessed a long journey over the last 15 years. Starting from very focused, specific entity/transaction monitoring to enterprise-wide monitoring to eventually AI-enabled smart solutions, IT has contributed and continues to contribute significantly to the monitoring of financial crime risk.

The current chapter opens with setting the context of financial crime committed through FIs in any customer lifecycle. It then gives an overview of various modules/products applicable in dealing with financial crimes, along with their key features. Customer reviews focused on their satisfaction and dissatisfaction drivers, are discussed. This helps all learners to understand the focus areas when they evaluate any product. The last section provides an overview of AML software markets. Special attention has been given to new solution features that are required for the sector and are launched by fintechs and regtechs. The chapter ends with a case for collaboration by regulators, financial institutions or product providers, to create a financial crime bureau for information exchange. These measures can significantly enhance the transaction and entity visibility of those who perpetrate financial crimes or are recognized as conducting suspicious activities.

Financial crimes in the context of the customer lifecycle.

A. Gupta et al., *Artificial Intelligence Applications in Banking and Financial Services*, Future of Business and Finance, https://doi.org/10.1007/978-981-99-2571-1_3

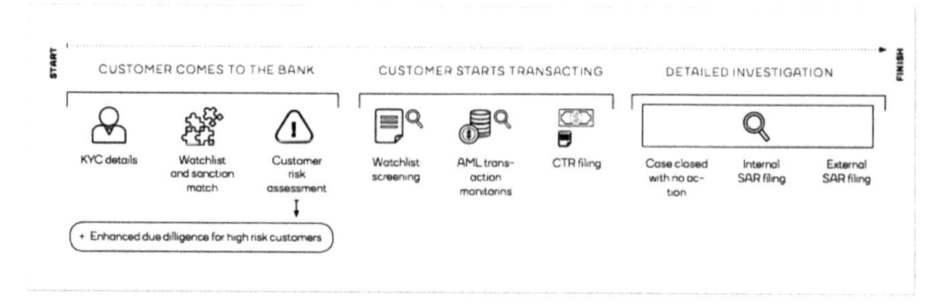

Fig. 3.1 FI actions during the customer lifecycle

Before understanding the solutions overview and capabilities, let us first understand the activities that a financial institution needs to perform to identify risky customers and transaction, and how they can be mitigated or detected.

As shown in Fig. 3.1, there are multiple touchpoints that are leveraged by the money launderers.

The first touchpoint is the opening of the account. It is the responsibility of a financial institution to identify sanctioned individuals or companies and to refuse to open an account for them. There are a few international agencies (OFAC, FBI, UN watchlist, and others) who provide lists of deemed unlawful entities. Banks and other FIs are required to match their potential customers against those lists.

The second touchpoint of exposure starts with the first transaction. It continues with further activity, until the FI identifies suspicious behavior and investigates it.

The third touchpoint is money transfers for terrorist financing and to sanctioned entities. Transferors are generally aware that their beneficiaries are unlawful entities. Hence, it is imperative for the FIs to detect such transfers and stop them.

It is important for an FI to have robust policies in place that serve to prevent financial crimes, rather than letting them happen and dealing with the consequences. That is why high-risk customers are not onboarded, and large cash withdrawals are permitted only after identification, ascertained through valid methods of proof of identity. However, the perpetrators are often well prepared and know how to bypass the proof of identity "barrier" and easily commit the crime.

The objective of an enterprise solution is threefold: (a) the ability to identify risky entities and individuals, who may perpetrate financial crimes; (b) the ability to track transaction patterns to identify anomalous or suspicious transactions. Such transactions are flagged and investigated to check for the actual intent of the customer; and (c) to monitor counterparties of the suspicious transactions to ensure that no banned entity, or any related party, can execute criminal transactions through legal channels.

3.2 Modules of the Solution

Enterprise financial crime control solutions typically have the following products or modules:

1. Customer onboarding solutions—KYC risk scoring
2. Sanctions screening and PEP matching for onboarding customer
3. Anti-money laundering (AML) solutions and real-time sanctions monitoring of financial transactions
4. Ongoing customer due diligence.

3.2.1 Customer Onboarding Solution—KYC Risk Scoring

The customer KYC questionnaire (know your customer questionnaire) is now an integral part of any new account opening process. Regulators across the world have mandated companies to collect standardized information about new customers before they start dealing with them. Financial institutions are also required to conduct their risk due diligence. Based on the customer risk assessment, the financial institution is expected to monitor high-risk customers more closely than their low-risk ones.

Historically, financial institutions were monitoring risk subjectively. Policies used to define riskiness on certain dimensions (e.g., high-risk country of origin). The bank team subjectively applied these criteria to classify a customer and conduct enhanced due diligence if required.

With time, many digitized solutions have been created to automate this manual process. Some of them are:

Digital KYC
The primary task of this product is to use machines to automate the manual tasks that otherwise a human would perform. For example, a popular digital KYC uses optical character recognition to read a scanned document and pre-fill the majority of the customer information automatically, without a need to manually feed in the form. The customer simply needs to review and verify the content.

Video Call Verification
A video call can now help to verify the individual's identity and complete the KYC process without requiring a physical meeting for document verification. It takes away a lot of inconvenience, boosts the productivity of the staff, and improves the customer experience.

Another key feature of the KYC solution is enabling users to create different logical workflows. For example, the solution should have the flexibility to first perform a National ID and Tax ID verification, through APIs. Only the successful matches should be forwarded for manual data entry or verification. Another option could be a conditional onboarding with limited information, subject to detailed verification, and so on. Each organization can have a different process, based on their business requirements. A good solution should have the flexibility to cater to the required workflow with minimal programing or customizations.

Sanctions screening and PEP matching.

Once the KYC information is collected and it is ascertained that a new customer is not a banned individual, the bank can move to the next step. The FI needs to know whether the customer is politically exposed (PEP—a politically exposed person).

The related parties of PEPs have shown a tendency to access illegally gotten, corrupt money on a large scale. Hence, their transactions are closely monitored, and all suspicious activity is reported.

How does an FI know that the customer is politically exposed? There are special lists of PEP individuals published by the international agencies that probe crimes of a financial nature (UN, FBI, US OFAC, EU, country-level regulators, Interpol, etc.).

Watchlist Aggregators

There are organizations that aggregate PEP watchlists and make them available to users through subscription. This can be helpful for the FI, as it minimizes the manual task of compiling the watchlist themselves, from over 10 to 15 different sources, on a regular basis.

Watchlist Enhancements

Another value addition by the watchlist aggregators is improving the extent of the details captured, on watchlist entities. Few companies specialize in such information enhancements (Reuters, Dow Jones, etc.). They conduct press searches and append more details about the sanctioned entity.

Naturally, richer information means fewer false positives in name matching. It is possible because of an improved fill rate on additional fields like country of birth, tax ID, company information, corporate structures, key stakeholder's information, etc. Otherwise, such details are generally not well captured in the original lists.

Some of the important features of the sanctions products are:

- Ability to connect to internal or external watchlists
- Ability to upload whitelists (names that should not generate alerts, despite their getting a match)
- Ability to provide exact and fuzzy matches for different accuracy levels
- Ability to generate a case for investigation
- Ability to provide operational and strategic reports to stakeholders.

Risk Assessment

The next stage is the risk assessment of the clean customers (those not on the watchlist).

The KYC information, along with the knowledge of the customer's banking needs and his potential banking products, forms the grounds for customer risk scoring. Based on the score, the financial institution can decide to execute a deeper investigation on the source of funds, intent of banking, details of turnover, etc. (this is called enhanced due diligence).

From a solution point of view, the risk assessment module takes care of the following:

1. A scorecard that can score the riskiness of a customer based on the provided input
2. A workflow to conduct enhanced due diligence and store the decision along with relevant documents
3. A reporting engine to provide stakeholders with the overview on operational and portfolio KPIs.

Depending on the sophistication, the scorecards can be expert based or statistics based. In the first case each factor and score weight is based on business judgment. For the statistical scorecards, however, the solution should be able to handle machine learning algorithms as well.

3.2.2 Financial Transaction Monitoring

The onboarded customer starts his banking activity. The challenge for a financial institution is to segregate his good transactions from the suspicious ones.

An effective transaction monitoring system can analyze all relevant aspects, including:

- Entity and individual-level interactions
- Mode of payments
- Transaction patterns
- Transaction values
- Structuring of transactions
- Source and destination locations.

These functions are typically organized and managed in the form of a scheme or a typology. Every such scheme is a business pattern that recognizes a potential money laundering event.

For example, one bank in Europe has typology for money laundering between Southeast Asia and South America for financing the drug cartels—through a European bank as a conduit. This typology needs to be auto monitored by the AML solutions in terms of the source and destination of the funds.

For all matching transactions, an alert will be raised automatically if they exceed a particular aggregate value (frequency of transaction, value of transaction, etc.) within a given time frame.

Once the alert is generated, the investigation team is required to deep dive into the event. Investigators obtain more details from the customer and conduct their own search to check if the transaction is genuine. Otherwise, it is classified as a suspicious event.

Based on the outcome, either a SAR is filed, and the bank exits the relationship with the customer, or the case is closed with no further action.

Keeping the above in mind, this module typically has the following features:

- Behavior analysis through the transaction patterns
- Set of rules for transaction monitoring. It should automatically capture events like high value transactions, etc.
- Aggregation of transaction values from different modes (wire, check, cash)
- Monitoring against transactions done with watchlist entities, high-risk countries, and high-risk counterparties.

Such monitoring enables financial institutions to identify customers who tend to conduct suspicious activity and weed out the money launderers among them.

The modules described above are made available to FIs, depending on their needs. But there are a few features that are provided to all companies, irrespective. They are.

3.2.3 Case Investigations

The majority of compliance solutions are alert-based systems. They observe an anomaly—be it a watchlist name match or a suspicious transaction behavior. Based on the anomaly, the system generates an alert for the team to investigate and take appropriate action. It is necessary to have a strong investigation module within all solutions.

Case/alert investigation modules enable users to:

(a) Open or create a case
(b) Add or modify multiple alerts into a single case, so that all related transactions and alerts can be investigated together
(c) Work and collaborate with peers if required
(d) Document the findings and upload supporting documents
(e) Create a workflow for an investigation process that can be systematized with central control
(f) Have a secure user authorization and access control to ensure the right authorities can observe and modify the cases as per their permissions.

These functionalities are a bare minimum for a smooth investigation. But with time, developers started focusing on automation and providing more features to enhance productivity for the investigator.

3.2.4 Reporting

The AML department and the compliance department should provide various reporting to their stakeholders. They are broadly classified into:

(a) Operational reports. Aging of cases and alerts, number of investigators, and average number of cases handled by investigators
(b) Scenario effectiveness. Number of alerts by scenario, the false positive rate, and scenario effectiveness in terms of productivity rates
(c) Strategic reports—total laundered money protected, number of money launderers by segment, scenario effectiveness, total operational cost, service-level parameters.

A solution should provide most of these reports out of box. Others should be configurable when needed.

Generation of SAR/CTR Report
In SAR or CTR report filing, it is important to **document the suspicious activities** that led to the report creation. Another important feature is storing the **narration** that prompted the institution to file for a SAR.

A manual exercise can take a lot of time. A good solution is the ability to automate the majority of the documentation, narration, and formatting of the document as per the regulatory requirements.

User Access Control
This module may be the last in the list, but it is not the least of it. All solutions, irrespective of their application, intend to have user access control, single sign-on capabilities. While the future of cyber security is now evolving to new concepts rather than session-level authorizations, the current solutions still rely on them. Access levels and rights are assigned to the users based on their defined capabilities. While access control is purely a technical product, it is defined as one of the industry's business requirements.

3.3 Backend or Technical Functionalities

A robust solution typically has both functional and technical capabilities. While business users of the solution are satisfied with the functional features, the robustness and security of it always depends on the technical (backend) capabilities—the solution architecture. Here are the most critical aspects of it.

Data Integration Capability
AML solutions work with both real-time and non-real-time data. The real-time mode is used during onboarding, financial transactions monitoring, watchlist name

matching, etc. Non-real time can be applied in identifying money laundering patterns.

Not only that, the solution itself consumes data from multiple sources: core banking solutions, KYC solutions, customer onboarding solutions, SWIFT, and others. A solution's flexibility to integrate across different platforms is another important attribute. If the systems are not flexible for seamless flow of data, it becomes difficult for the financial institutions to customize, add, delete and modify the sources flowing into the solution.

IT Security of the Solution
Any financial institution will typically have a security checklist with 60–200 parameters that include framework for single sign-on, browser support, security guidelines applicable for the organization, and similar requirements. A solution is supposed to fulfill all functions, such organizational needs.

Audit Trail
The last of the key features of the product is the audit trail. Ideally, all actions (logins, modifications) should be stored in a log automatically.

These components, both functional and technical, are considered to be "must-haves" from an AML solution suite perspective. In the next section, we will focus on organization needs that go beyond the bare minimum.

3.4 Organizations' Needs from AML Solutions

Many developers design custom solutions to address particular AML-related needs. Some of them create value in a specific component and some provide an enterprise-wide capability.

The section below is based on our team's research. We interacted with the clients, analyzed client reviews, and used our own experiences in consulting projects to identify the actual requirements for the AML solutions. Figures 3.2 and 3.3 are outcome of internal team's research.

From a functionality point of view, a good AML solution should have following characteristics.

A Scalable Solution
The solution should be able to fulfill all the regulatory requirements of the country where it operates. Different organizations, depending on the nature of their business, customers, and geographies have varying risk classes and asset class exposures, for different lines of business. The solution should be able to cater to the needs of such diversity without complex customizations.

Ability to Provide Interface to Front, Mid and Back End System
Depending on an organization's process, the solution should be easily integrated with the interface of front, mid of back-office systems. All required fields/reports should be readily available to the different stakeholders across the process.

Fig. 3.2 User review on the strength of their existing AML solution ($n = 52$)

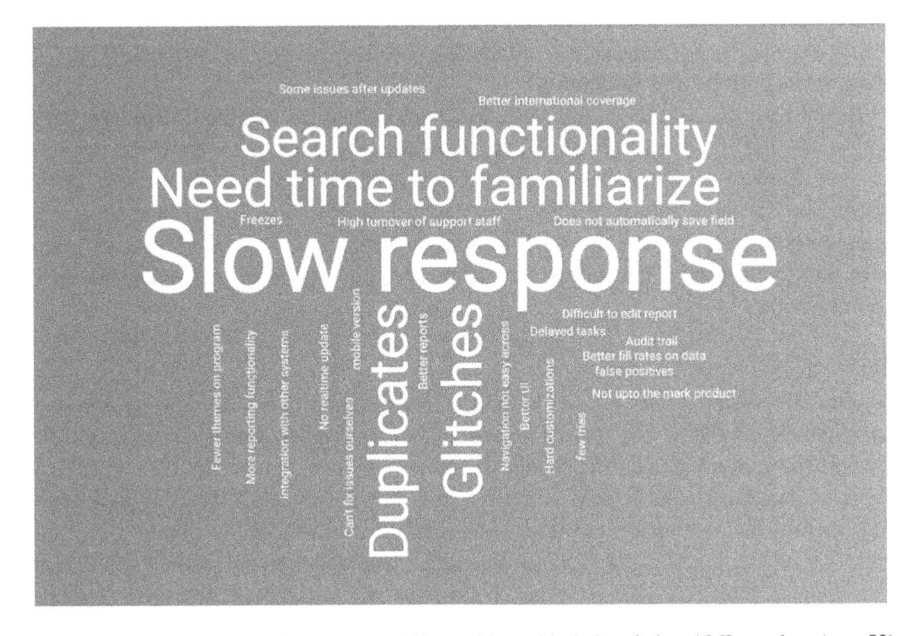

Fig. 3.3 User review on their concerns while working with their existing AML product ($n = 52$)

User Friendliness
Clients prefer solutions with simple and intuitive interfaces, easy to follow processes, and simple user boxes. One organization selected a product based on the number of clicks that an investigator needs to do to open the case. The company's management mentioned that one less click means a few hundred hours of the investigator's time saved during the year.

Flexibility of the Solution
Many organizations have their own specific needs. It can be a need for creating customized scenarios, the addition of new business units, bifurcation of reporting for different units, or investigation workflows with customizable presentation of information. Some also need the capability to manually initiate a case, the ability to follow-up with other team members during an investigation or the presence of audit trail of all activities.

The more flexibility the solution can provide, the better it is. While the majority of these activities can be handled through customizations, the ease of such modifications matters. Complex customizations can become as costly and time consuming as the original solution implementation. Business teams and IT teams should keep this in mind while evaluating a solution provider.

Ability to Finetune the Thresholds for Transaction Monitoring
This has generally been a regulatory requirement as well as a business prudence. The portfolio of any organization is dynamic and so is the transaction behavior of the customers. It becomes important for the transaction monitoring system to redefine itself to continue being effective for the evolving environment. This is done through threshold finetuning.

Every rule that monitors a transaction behavior has a threshold. For example, if the rule looks for an excessive cash deposit, then the threshold value triggers an alert. Now consider a situation with rising inflation to buy the same amount of goods and services people will be using larger amounts of money. The rule will remain unchanged, but the threshold should be finetuned on an ongoing basis. This is in conjunction with regulatory guidelines that the financial institution needs to comply with.

Ability to Whitelist Customers
Another important feature is whitelisting of the customers. Take an example: a businessman in Delhi has started conducting a lot higher transactions of cash in the last three weeks, resulting in alerts being triggered against him for high cash value transactions. During the investigation, it was revealed that the individual was getting his property renovated. He provided a few bills but claimed that the majority of the contractors deal in cash. The investigator was satisfied with the explanation. However, he needs to ensure that this case does not keep flagging different investigators. The solution should be the ability to temporarily whitelist such customers with an explanation captured. Alternately, banks can upload a list of whitelisted clients through a backend process.

Ability to Upload FIs Internal Blacklist/Watchlist
Financial institutions also acquire locally prescribed lists. In some of the cases, they want to block blacklisted customers alongside the watchlist check during onboarding. This ensures that the bank is able to automate its screening process at one go. Keeping that in mind, another requirement from the screening solution is the ability to upload and manage their own defined watchlist or blacklists. A maker checker functionality here is also good to have.

Flexibility and Capability of Reporting Tools
As mentioned earlier, different stakeholders need to receive reports in different formats. A flexible solution should be able to generate these various reports. Reports should also be provided as per user access rights. Lastly, the reports should be accessible from different devices—desktop, notepads, android, mac, etc.

On-Demand Analytics
The customers finds self-help tools very useful. They want to be able to explore the trends and make minor finetuning to handle evolving tendencies. Hence, a solution that provides on-demand analytics will be in demand.

Efforts in Future Migration and Updates
Developers of the solutions keep upgrading them. For many developers, these upgrades become a forcing mechanism for generating ongoing revenues from the clients. It is good for the FIs to ensure that their software vendor is not going to overcharge for the future migrations and upgrades.

Big Data/In-Memory Data
Solutions are expected to be compatible across multiple platforms to accommodate the different sizes of organizations. We have been advising clients to leverage Postgres to manage the database licensing cost. We also suggest that clients adopt Apache Spark to take care of a large volume of real-time transactions in a matter of milliseconds. Solutions should have these capabilities or the vendor should be willing to provide the platform based on the customer's needs.

Open Source Platforms Friendly
A lot of organizations have started working on open-source machine learning platforms like R. Python. In many cases, FIs already have some existing functionalities implemented in these platforms. It can be models for reducing the false positive alert rate, or match rates capabilities. It is increasingly becoming important for the solutions to provide cross-platform support. No organization will want to forego their existing analytics initiatives.

Model-Ops and Cloud-Based Operations
Migration to managed cloud is becoming a new trend in Southeast Asia, Far East Europe, and going up to North America. Even in Middle Eastern markets, non-banking service companies have started adopting managed cloud. The ability to

work on Kubernetes or native cloud is becoming a normal requirement for the solutions. We expect that with the eventual application of machine learning and AI, the organizations will gradually migrate to model-ops and dev-ops for their compliance functions, as well.

More Sophistication to Optimize False Positives
This is now a play area of the fintech and regtech startups. They are bringing innovation and smarter solutions as plug-ins to reduce high false positives. They are used in sanctions screening, transaction monitoring, and case investigation time optimization. With the success of these startups, large organizations have also started slowly launching AI-enabled AML solutions.

Rise of Artificial Intelligence and Machine Learning in AML Software
As the complexity of money laundering is increasing, the pressure on the department also keeps growing exponentially. The regulatory penalties, regular central bank audits, and reputation risks are so severe that it has forced some of banks to exit certain markets or customer segments. We are aware of quite a few financial institutions that revised their approved annual budget for higher alert investigation—just because the central bank induced new AML areas where banks were not tracking suspicious activities before.

This has put a lot of pressure on the financial institutions to deliver as per expectations without increasing cost and without putting significant pressure on their teams. This is where organizations are moving to artificial intelligence and automation (Fig. 3.4).

3.5 Market Overview of AML

The size of the AML market is almost USD 2 Billion, and it is expected to continue growing at a healthy rate of 20% over the next five years (Fig. 3.5).

As shown in the graph above, some of the major drivers are increasing money laundering and terrorist financing activities, that are prompting the regulatory agencies to be more stringent and issue new guidelines and enforce audits.

This growth is driven by the following segments:

- **Size of companies**. The small and medium-sized enterprises (SMEs) segment is expected to witness the highest growth. SMEs are increasingly being pushed across different sectors. In many countries, jewelers, fintechs, brokers, exchanges, etc. also have to implement anti-money laundering and terrorist financing monitoring solutions. Their limited budget keeps this segment open to the cost effectiveness of AML products. They are also expected to look for cloud-based services faster than on-premises.
- **Geographies**. There are two geographies that will drive the future growth—North America will remain the largest market from a revenue point of view. The rising regulatory oversight on money laundering and terrorist financing

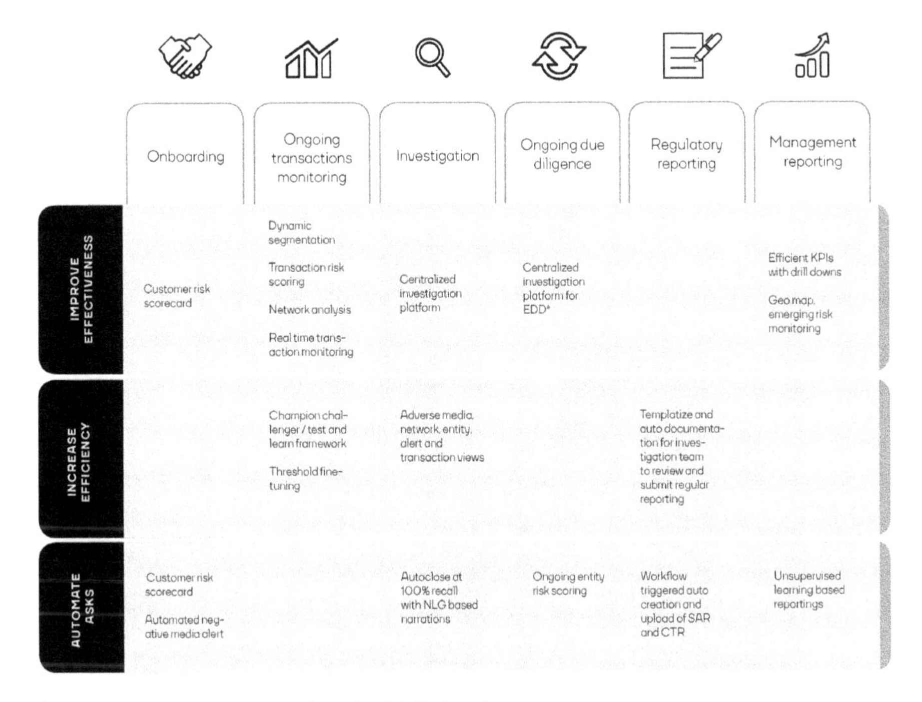

Fig. 3.4 Digitization opportunities for AML function

activities, rising e-channel transactions are drivers for the North American markets. Asia will be the second most important market. In that locality, it is driven by increasing regulatory pressure, rise of e-payments, fintechs, SME, and non-banking financial companies (NBFCs). As apparent from the drivers, this segment will also be price sensitive.

With the pie of opportunity increasing, the competition is heating up. There were an estimated 140 companies that offer enterprise solutions, ID match software, fintech, regtech, and KYC/CDD solutions, in 2018. This number has increased to approximately 500 + companies dealing in this space. The majority of them are still small enterprises, with a focus on niche areas. The biggest addition is driven by regtechs and fintechs that have tried to leverage machine learning or digitization in the AML space to provide innovative solutions.

A quick breakdown of global companies as per our research is as Fig. 3.6.

Analysis suggests that the fastest rise has been witnessed in ID verification, and watchlist and PEP aggregators.

The Enterprise AML, or financial transaction monitoring offerings, have increased not only in terms of number of providers, but most of the new entrants are trying to leverage AI as a play to gain customers of existing software providers.

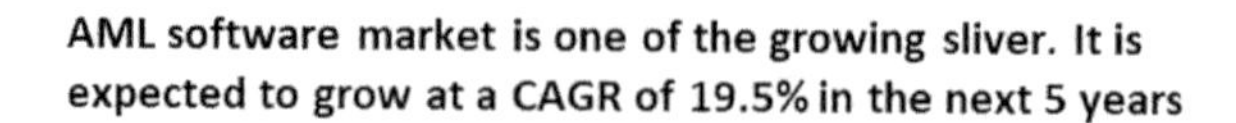

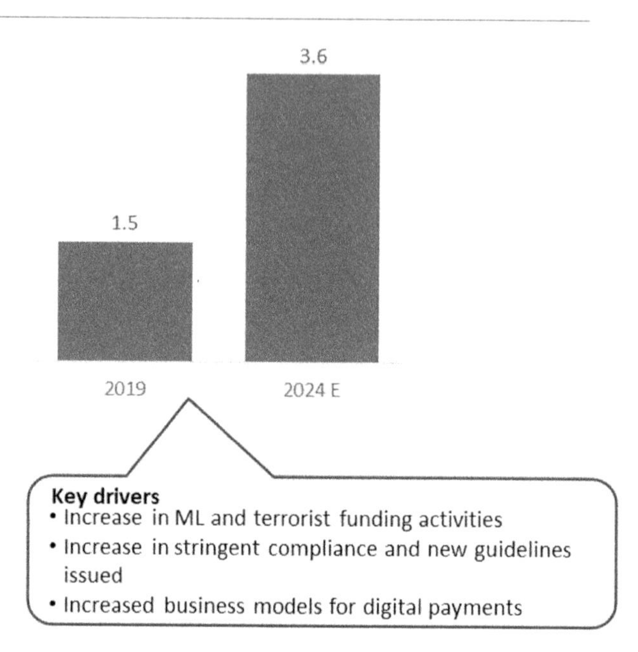

Fig. 3.5 Market size in USD billion and expected growth of AML solution. *Source* Research reports, team analysis

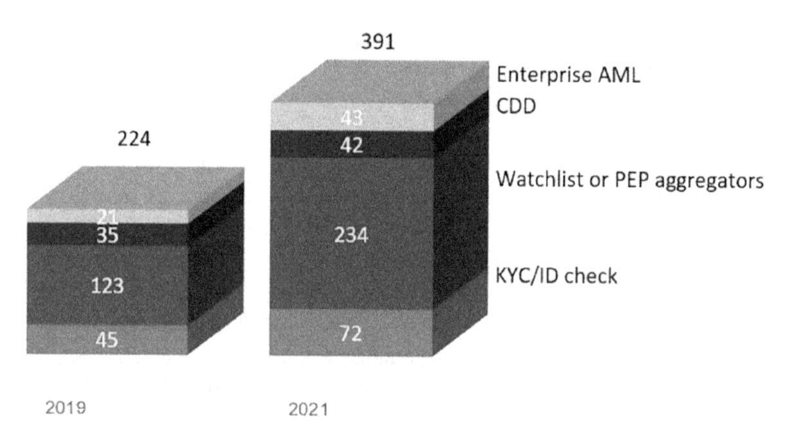

Fig. 3.6 Number of companies operating in financial crimes monitoring space

3.6 Emerging Trends in AML Solutions

Modern AML solutions are diverse because the needs of the companies change from organization to organization. Their main emerging needs are:

Capability Suite Rather than Modular Capabilities
Companies do not want to deal with multiple providers for different functionalities. They are generally happy with a provider who can offer all capabilities rather than integrating and managing multiple functionalities.

Easy to Customize
Solutions that are easy to customize based on the evolving needs are becoming more popular.

Advanced Analytical Capabilities
Companies expect providers to embed advanced analytics, machine learning and automation to better help their investigation team. They look for smart yet cost effective solutions.

NLP Capabilities
Some of the advanced FIs expect solutions to mine unstructured text, case narrations, news and many other textual information. They want it to be summarized and be readily available for identification or during investigations.

Network Analysis
Few innovative FIs want to analyze not only their customers' transaction behavior, but also customers' networks and transactions of their counterparties. It helps them to track suspicious entities and organizations more effectively.

Collaboration
Like credit risk and fraud management, there exist opportunities for institutions at different levels—cross-country level, intra-country level, different industry level, to aggregate and share the information for effective monitoring. Information on high volume transactors and money launderers can be collected across organizations on a regional level. This can be a game changer, as banks will get a clear visibility on transactions, including those outside their own network. Creation of a financial crime bureau (along the lines of credit bureaus) would be a revolution in the way organizations manage their anti-financial crimes practice.

This will also prompt regtechs to generate newer solutions to provide smarter, automated offerings.

We are hoping to see more advancements in AML solutions in the next five years.

4 Data Organization for an FCC Unit

4.1 Introduction

Data is one of the most critical elements of an AI organization, irrespective of the department or function it serves. This has been in the knowledge of almost all business folks across the world for some time now. However, dedicated focus to get the maximum out of it has been a work in progress.

Most of the large organizations are not able to leverage the value of machine learning (ML) and artificial intelligence due to the lack of good quality data. The ability to identify the quality of data fields that are important to providing good insights from AML perspective is a critical success factor.

Key questions related to data governance are generally pursued organization-wide rather than being handled as a responsibility of FCC unit. Sometimes data fields that are present for risk management and fraud control units are not available to the AML teams. The authors, through this chapter, will lay out key dimensions of data that is required for a best-in-class FCC organization. We will also highlight issues like information captured on a suspicious activity report (SAR) filing and the timing of case closure that can have a significant bearing on the quality of insights and the models.

This chapter will also discuss data access and transfer issues and the potential challenges they pose to multinational enterprises, especially the cross-border flow of data.

Finally, we will lay out a conceptual enterprise-wide data organization and the key principles of data governance that should be implemented for a high-quality ML-AI-driven FCC management establishment.

The discussion on artificial intelligence must logically start from data. Without good quality data, the conceptualization of an intelligent solution is meaningless. Despite this common notion, data has been one of the biggest differentiators for successful AI organizations and a showstopper for quite a few organizations, where AI initiatives in compliance, including in AML have not delivered desired results.

A. Gupta et al., *Artificial Intelligence Applications in Banking and Financial Services*, Future of Business and Finance, https://doi.org/10.1007/978-981-99-2571-1_4

Before delving deeper into the data, let us first understand a few organizational drivers that need to be tackled first.

Believe it or not, it is data and not the business that can define the success of intelligent AI programs. As consultants, we have often come across financial institutions, where business is very keen on defining their expectations from AI. They are very articulate on the pain areas and how AI should translate into use cases for them. However, when it comes to data-related challenges, businesses are often seen to "outsource" that activity to their IT departments. The approach becomes a big differentiator of the future success of AI programs.

As shown Fig. 4.1, there are different types of organizations, based on the involvement of the compliance department. Best-in-class organizations will have compliance departments actively engaged in the potential benefits of AI and working with IT in identifying data-related issues. The next stage is a joint vision in terms of road map on data infrastructure that is co-owned by compliance and IT. This is an ideal combination for a successful organization.

We would also highlight that there exist diminishing returns in improving data infrastructure. However, a big value of the solution intelligence can be delivered if compliance department is engaged and works closely with IT.

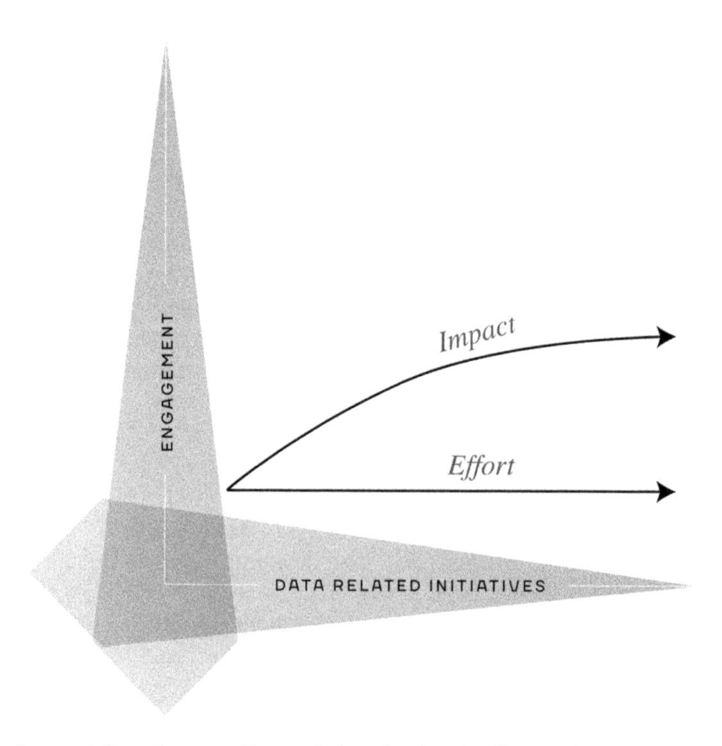

Fig. 4.1 Impact delivered versus effort evolution showing significant gains during initial focus on data initiatives

Let us try to explain it through an example. In one of our recent engagements for developing a machine learning-driven alert scoring model, we realized that the individual customer segment has customers whose age is approximately two years. After discussion with the business, it was clear that minors cannot have accounts in the country. A deeper investigation revealed issues with the classification of customers into individual and non-individual segments.

The recommendation from the consultant was to revisit and improve the definition of segments. The reason for consultants to push this change was that for a specific segment, time since incorporation (age of company) was a major driver to suspected money laundering activities. The compliance department, while agreeing to the observation, parked this issue, as it has a broader implication. Their alert generation process was leveraging this segmentation. Their investigation teams were used to seeing a segmentation in a particular way. Business finally asked the consultants to continue with a slight dip in model performance, but no change to segmentation. Not even an agreement to revisit it, in their next year's roadmap. There are more such data issues, that consultant identified. However, many of them are parked due to a lack of support from the compliance department to sponsor a data governance project as a part of the AI journey.

Another example was related to data integrity. During data quality assessment, the team identified that there are customers, whose account open date is later than their first transaction date. It clearly suggests that the account open date has data integrity issues. The compliance team of another bank observed this issue and left it for IT to investigate and resolve. Eventually, the data team could not help. The consultant convinced all stakeholders to use the earliest account open date or the first transaction date as the account open date. The observation was for a non-negligible set of customers, hence the need for getting this cleaned. Eventually, we ended up with compromised data quality. In this chapter, we would like to explain the mechanism of data quality assessment. Our observations on a few specific issues of data quality have a high impact on the success of an AI-driven FCC organization.

4.1.1 Data Quality

Data quality for developing machine learning has four important dimensions:

(i) Data fill rates
(ii) Data integrity
(iii) Presence of outliers
(iv) Data history.

(i) Data fill rate

This pertains to the usability of data because of the presence of missing value within the dataset. Imagine generating insights from a dataset of 100,000 rows with only 5000 of them having valid values. Any inference drawn from only 5000 rows cannot be applied on entire dataset, as the inference can completely change with the increase in fill rate. The modeler tests this with a term called fill rate. Fill rates refer to the percent of a dataset containing valid values. Acceptable fill rates can change depending on the type of data for which the fill rate is calculated. The rule of thumb for datasets is typically 70%; i.e., if the fill rate is less than 70%, teams do not try to either impute missing value or consider the variable for meaningful analysis. However, there could be exceptions. For example, the number of customers is carrying traveler check from the bank. Not all customers of the bank travel abroad; even among the ones who travel, traveler's check is not very popular. This implies that if we were to analyze usage of travelers check in AML context, a fill rate of even 5–10% is meaningful and should be used.

Ways to assess fill rate:

Modelers typically analyze the percentage of missing for numerical variables. There could also be cases, where a modeler might consider three types of values as missing—commonsensical missing values, 0s, and special values like 999. The third one needs special attention. There are a lot of datasets, which are migrated from older systems. Those older systems have poor-quality data. During migration, IT teams replaced missing or junk values with special values like 999. They still mean missing, but technically are numeric. All the above types of values warrant investigation.

For character variables, modelers generate frequency distribution of character variables. Frequency distributions not only generate insights about the underlying behavior of a variable in a bank's portfolio, they can also generate values like "", ".". All of these would be referred to as missing and need treatment.

Figure 4.2 **showing summary of data quality assessment**

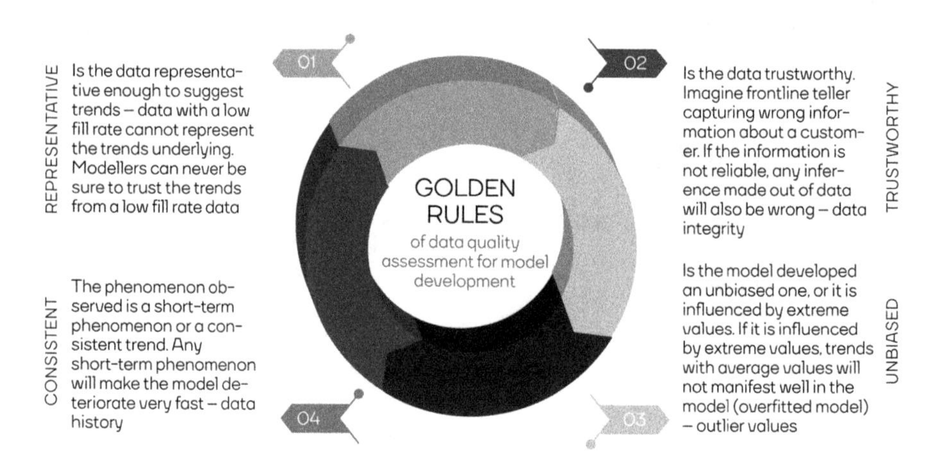

Fig. 4.2 Showing key elements of data quality assessment

(i) Missing value treatment

Before treating the missing values, we need to understand their nature. They are typically classified as:

- Missing at random
- Not missing at random.

As the name suggests missing at random means that missing values are randomly available throughout the data series; i.e., there is no pattern underlying. Not missing at random means that there exist some patterns in the missing values; for example, the missing rate is higher for certain branches or for a certain segment of customers. Depending on the type of missing values, the missing imputation techniques are adopted.

Some of the common missing imputation techniques are neighborhood imputation (median or mean). In this imputation technique, the neighbors of the missing value are considered and their mean or median is calculated to be used for imputation. Definition of neighbor can be modified to a completely random neighbor or neighbors organized by certain segment types. Another method for missing value imputation is the regression-based imputation. In these cases, a regression model is developed which predicts the values for missing value through some of the other variables which have much higher fill rate. This method is generally utilized for getting the best possible results; however, it is more tedious and sometimes impractical as missing values are common across different variables.

(ii) Data integrity

This is the next important assessment that is to be done. Even if data does not contain missing values, there could be data that does not really represent the true essence of the variables. For example, during a recent analysis of the pension fund dataset, we came across two variables. In the first one—employer type—the value for a big chunk of employers was the "private employer". However, another variable titled industry had a value "Ministry of HR and localization". During the investigation, it was told that a lot of small private companies do not handle the pension fund contribution themselves. They outsource it to a federal appointed agency, which in turn processes the contributions. While it is a meaningful insight for analyzing the business, it made the variable "industry" meaningless as almost 50% + values within the variable couldn't be inferred.

ML team should aim for a decent business understanding of the variables, as the level of insights and explainability of models depends on data and its correctness.

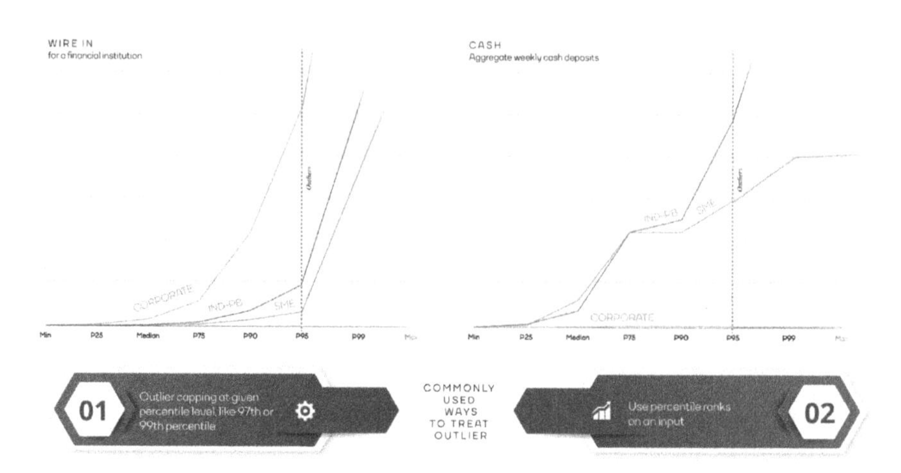

Fig. 4.3 Showing outliers and ways of treating the outlier

4.1.2 Presence of Outliers

Outliers refer to the presence of extreme values in the datasets. There exist variables like income and turnover of business that can have extreme values. Our experience suggests that most of such variables have the last one percentile of values as extreme values which are in multiples of the previous percentile values. While these values are valid, the challenge is that they can influence the model outcome significantly. For this reason, they need to be treated. There are many ways in which these outliers can be fixed. One can perform capping—defining a range beyond which these values can be capped. For example, the income beyond 1 million is capped at 1 million. Another way of defining capping is by defining capped values as percentile values. An alternative mechanism for handling outliers is to convert the original values into their percentiles ranks. Use the percentile ranked value as input into the model (Fig. 4.3).

4.1.3 Data History

Data history assessment is another important dimension to check before any data analysis is attempted. It refers to the depth of data in terms of historical time series available. It is important for us to understand the attributes and availability of the historical time series. If this information is available for a very short time frame, then it becomes very difficult to derive any meaningful inferences out of such short time series of data due to fewer number of dependent variables (SAR filing or serious suspicion resulting in case creation). Lesser number will mean compromised reliability of the model.

Secondly, the objective of any predictive or forecasting model is for the machine to be able to predict the future outcome with a certain degree of precision. There exist patterns like seasonality or structural shifts in the environment. A shorter time series dataset won't even be able to point issues. It could mean that models will deteriorate fast, if data history is not sufficient.

4.2 Data Dimensions Relevant for Analysis on Financial Crimes

To assess customer-induced financial crimes, a modeler needs to create a 360-degree view of the customer. This view comprises all information pertaining to the customer's demographics, banking behavior, transactions, information compiled at the time of KYC, and regular updates including CDD scores. Figure 4.4 summarizes various customer dimensions of data that are typically used for conducting data-driven analytics on the customer demographics and its transaction behavior and assess customer's riskiness for committing financial crimes.

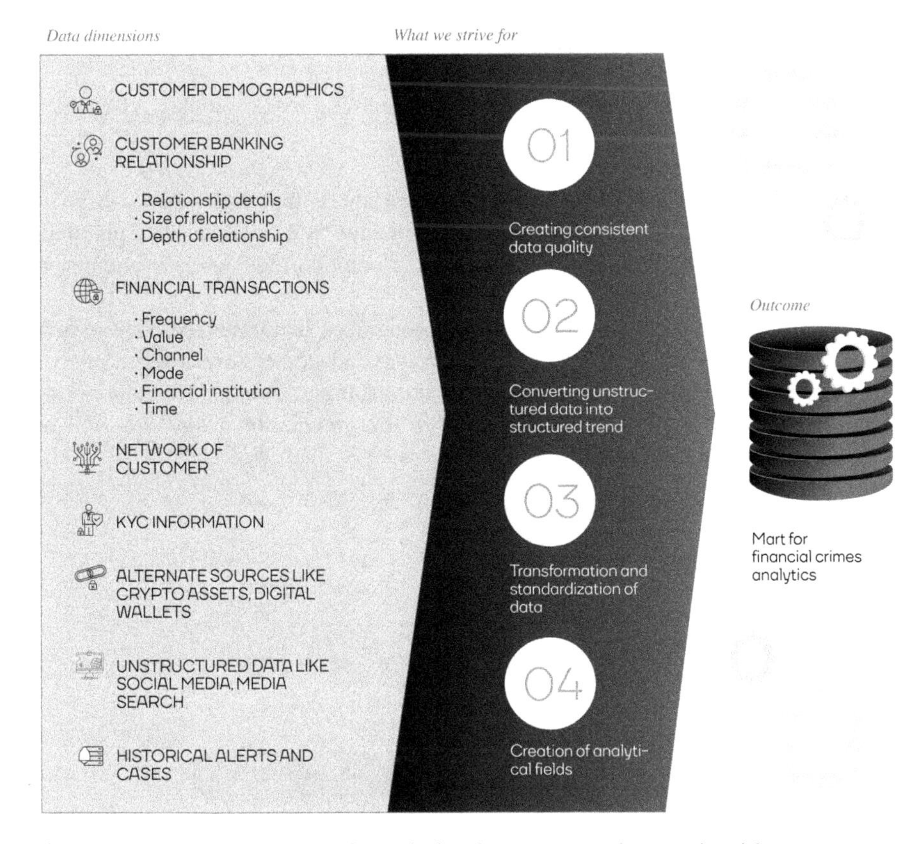

Fig. 4.4 Dimensions of data used for analyzing the customer and transaction risk

After the data quality assessment, data scientist starts creating derived variables. A customer's behavior is not reflected automatically through variables. Certain deep learning techniques like neural networks can be trained by providing raw inputs of transactions along with mode, date/time and other information. However, it is advisable to create some trends that are used to capture the behavior. For example, velocity of transactions, ticket size of transaction, counterparties information, background of the customer, and so on. A detailed list of such dimensions is provided in the appendix.

Another thing to keep in mind is that financial crime committers do not stick to one specific pattern. With the advent of new channels and payment solutions, the patterns of transactions are ever-evolving. This also necessitates a regular refresh of machine learning algorithms. However, the frequency and variables used for codification of such patterns also need to vary. For example, whether the customer will structure transactions within one day, one week, or one month, no one can precisely predict. Similarly, no one can predict whether most of the financial crimes will follow a particular type of pattern (in fact it is expected to be varying), and hence variables capturing transaction behavior in terms of velocity, ticket size, counterparties, etc., must be created with daily patterns, weekly patterns, monthly patterns, and other similar variables.

These variables will be fed into the machine learning for developing AI-enabled algorithm.

Data transformation

Once the data scientist has designed its variables, the next step is to write extract, transform, and load (ETL) queries. Typically, SQL, python, data pipelines or SAS programs are written depending on the solution that the data scientist is leveraging.

A typical data model for such a model mart is a denormalized view where every customer has derived fields organized at the customer level (depending on the modeling requirements, this can change. For example, if someone is attempting a model for transaction-level prediction, then the granularity of the row will be transaction and not customer) (Fig. 4.5).

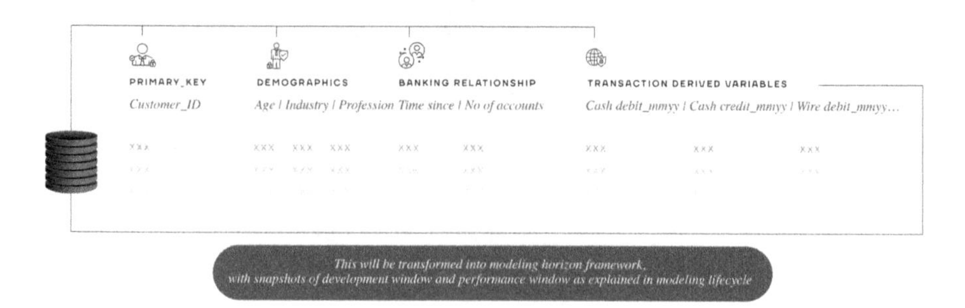

Fig. 4.5 Showing denormalized view for conducting analysis and model development

The transformation of raw data into a denormalized form is called the process of modeling mart creation. Every solution has provided a nomenclature to it; however, generically it is a model mart that is used for training the machine learning models.

4.3 Cross-Border Data-Related Challenges

Cross-border data has its own challenges driven by both the regulations and the multiplicity of the agencies that compile and consolidate the data. In fact, there is an ongoing exercise in Europe that is going through these challenges. As a part of ultimate beneficiary library, all countries are compiling information related to the ultimate beneficiary for every entity. This needs to be an EU-level information sharing. It will facilitate a clearer view of anyone who is layering money internationally to launder money. However, the countries are facing challenges as the codification of information across countries is not standardized. Similarly, the source for the data collection is also diverse, depending on the country. This puts a question mark on their integrity and uniformity. These cross-border datasets will always need standardizations in terms of definitions and agreement on the reliability of the source, from where the data is compiled.

One of the challenges that a data scientist face is handling cross-border situations. A lot of financial institutions are multinational. While there exists specificity in every country, centralized analytics might not be feasible. The reason for this is that the local regulations do not allow for sharing of data to their Head Office (HO), even for analytics, consolidation of information for reporting, or for developing machine learning models.

Another challenge with cross-border information is that the depth of data varies across countries. There will be countries where customer acquisition is rich in the high-net-worth individuals (HNI) segment, but mass customers are not many. There are countries where small and medium enterprises (SMEs) are many, but corporates and consumer banking customers are very few.

This poses unique challenges for data scientists. We are often faced with a situation, where the financial institution wants to do threshold finetuning of a few alerts for a specific country, but the number of transactions and alerts to analyze and assess productivity and underlying transaction distribution cannot be done, because of a lack of data depth.

Secondly, compliance teams in a lot of countries restrict access to demographic information that leaves the data scientist with minimal information for conducting meaningful analysis.

Data scientists often find ways to tackle these challenges without compromising the quality of analysis. As consultants, some of the mechanisms adopted are:

Creating synthetic data

If there exists a portfolio that does not have depth on data, or we are faced with a new portfolio, one of the mechanisms is the creation of synthetic data. The process of creation of synthetic data is explained as follows:

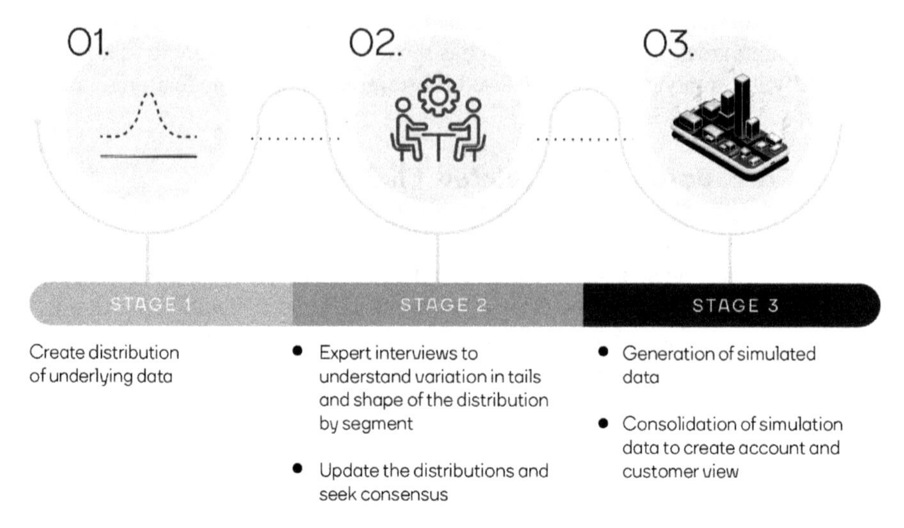

Fig. 4.6 Showing the process of developing synthetic data

As explained in Fig. 4.6, the first stage entails the creation of the distribution of the underlying portfolio. Typically, this is generated for a different mode of transactions and by segments.

Stage 2—Adjustment to the distribution

In this stage, data scientists need to make an adjustment to the distributions. For example, let us look at the example below for cash transactions of an HNI customer.

Say the data scientist realizes that the HNI segment in this portfolio is not doing large transactions. Based on the experience of portfolios in similar countries and expert opinions, these can be modified.

Stage 3—Generation of simulated data

Once the distributions are generated, the last stage entails simulation of data for an account and eventually rolling up to customers.

Data generated through this can have the necessary number of observations to generate meaningful analysis.

Similarly, SAR-filed customers data can be generated

Expert-based synthetic data

In this case, step 1 of the process is not done. The distributions on transaction types by segment are generated based on expert inputs. The data scientist can also adopt consensus estimations by using the Delphi technique.

4.4 GDPR-Related Data Challenges

European Union introduced General Data Protection Regulation (GDPR) in the year 2018. Before this regulation, the organizations were already aware of the data privacy of the consumer. The institutions were also designing and implementing data privacy policies for their organizations; however, there were also cases of private information used by business. GDPR has provided a regulatory framework for the organizations across to follow a data privacy and usage of their customer data. The act aims to:

- Provide transparency and upfront disclosure to the customer about the fact that their personal data is being captured and will be used by organization
- Ensuring security of customer data through stronger data governance approach as well as providing security layers to the data
- Have a clear policy around data sharing and its usage by type of data. It ensures minimal misuses of customer data collected for desired purposes
- Documentation of the processing of personal data
- Ensuring that the data breaches are not hidden. In fact, they are reported within 72 h of such incidences happening.

GDPR was brought with the right intent of providing security and purpose to privacy of customer data and, however, has also created its own downside in the regulator's efforts on creating an exercise of compiling the ultimate beneficial owners.

This exercise was conducted, keeping in mind the rising instances of cross-border layering and structuring of laundered money. Unless the beneficiaries across jurisdictions are captured, standardized, and then exposed to the financial intelligence units across countries, individual jurisdictions will always have a limited view of the transactions and the behaviors.

There are two major challenges that are currently being faced by the regulators:

1. Differing levels of data quality and convention due to several third-party data aggregators being used in different countries. Even their reliability and verification process also vary
2. With GDPR regulation applicable EU-wide, there exists hesitancy in sharing the information.

EU's efforts could go a long way in defining a framework for other regulators to collaborate and unearth a lot of cross-border layering, by simply making the information standardized and easily accessible.

Initiatives like these need to have a clearer direction through regulators across the globe for specified purposes. Otherwise, cross-border data collaboration will become an impediment to increasingly complex transaction trails that money launderers are following.

While technology solutions can be generated that enable only machines to view personal data and take business decisions or summarize it for the users. However, harmonization and agreeing on common and verifiable processes to enable machines to leverage this is still going to be a slightly long-term plan.

Implications for An Organization
For an organization to be compliant with GDPR, it needs to do the following:

(i) Create a robust data governance policy
(ii) Create sound processes and controls for managing consumer data
(iii) Invest in technology for data protection and encryption
(iv) Leverage artificial intelligence (AI) and other digital tools for automation of tasks that are to be done on a regular basis.

(i) Create a robust data governance policy

A robust data governance policy is a requirement irrespective of GDPR or not. Readers will note this in next section on "areas of improvements". Defining the source of data, assessing the quality of data, ensuring data steward roles, ensuring IT security and data security policies and proper dissemination of those, ensuring encryption of data and security protocols, and finally auditability of data accessed. Lastly, a proper data privacy policy which complies with GDPR is also a must have.

We would not delve deeper into these topics as there are many books on these topics but ensuring that a compliance organization has a robust data governance policy is very important.

(ii) Create sound processes and controls for managing consumer data

This is another important aspect with some overlap on data governance. An organization needs to clearly define the personal information of consumers. The data privacy policy should be transparent and easy to understand by the consumer. They can then provide consent to the usage of information.

Financial institutions can define the data journey for different sources in their enterprise-level architecture (whether on cloud or on-premises) and then map the above to ensure that the bank has a scalable and flexible framework for data management. It also lays out guidelines for future updates. The data cataloging and glossaries along with their owners will ensure that there exists a well-defined process for managing and updating data.

Another important aspect is to ensure that IT or data teams have enabled auditability at the field level for an end-to-end data journey.

(iii) Invest in technology for data protection and encryption

Data protection is an important aspect of data governance. Aligned with a data privacy policy, every bank would need to clearly define data that needs encryption, data that needs anonymization, or pseudonymization. Another important aspect of this would be defining roles, responsibilities, and access rights for organizations and partners including outsourced entities.

(iv) Leverage artificial intelligence (AI) and other digital tools for automation of tasks that are to be done on a regular basis

Lastly, once the data journeys, enterprise data governance, and policy are getting settled, the next step would be to leverage technology. Organizations can map the processes, e.g., data processing of consumer data, identification, and applying data policy on consumer private data through AI, audit reports, and automated data quality reports. Model-ops-driven data-driven insights lifecycle. There can be many such use cases that can be pursued.

4.5 Areas of Improvement for Creating Best-in-Class Data Organization

Often artificial intelligence programs failed, more so in compliance departments. The biggest impact areas for data-driven insights from the perspective of a financial crime are the following.

4.6 Knowledge of AI and Its Enablers for Compliance Heads

One of the common themes that authors witnessed is that majority of the head of compliances are aware of the application of AI in their business. Many of the heads are exposed to use cases in their departments. Despite the varying degree of knowledge, all compliance heads have some sense of business applications where AI or bots can help in their business. The majority have explored a few bots on investigations. However, when it comes to understanding the lifecycle of data to machine learning and their translation into business impact, their knowledge is minimal. The reason, we believe that it is dangerous is because businesses are not seen getting their hands dirty on issues related to data. IT and data teams in a lot of organizations are not opening another door for their enterprise data. A big chunk of emerging and a few institutions in developed markets are creating another duplicate enterprise data platform for compliance. Given the domain of compliance is new, data infrastructure for compliance has the same problems that were witnessed for risk management 10 years back. There exists an adhocism to data exercise. Business doesn't want to get their hands dirty to understand and resolve data quality issues.

The second dangerous zone for the users is that neither the head of compliance nor anyone in their teams is an experienced, data scientist. This means that assessment of the quality of models, initial pilots to slowly understand the impacts of machine learning and then rollouts, etc., are faced with an impatient audience, and sometimes running fast is resulting in either failed projects or in a few other cases much lower impact resulting in disappointment of stakeholders toward AI.

4.7 Improving Data Quality and Integrity in KYC

While working on financial crimes engagements, our experience has not been great on data quality for KYC information. Be it defining the fill rate for KYC customers or the integrity of data captured; frontline does not do a great job in capturing the information of the customer. The lack of enforcement on data quality here also doesn't help.

4.8 Desiloing of Data

Another important aspect of best-in-class data foundation is the desiloing of data. Financial crime analysis and investigation require customer data. For example, almost all investigation professionals said that they rely on KYC data during the investigation to map transactions against customer profiles. Despite this commonly stated argument, KYC data is always extracted manually. Similarly, loan prepayments, credit card-related information, investments, and assets under management data are generally not consolidated and provided to the AML stakeholders. Providing a comprehensive 360-degree view across assets, demography, banking relationship, transactions, and counterparties is an important step toward creating a world-class data organization.

4.9 Having the Right Lead for Data Organization

The data journey itself is a long process. For an efficient organization, it can be of 6–9 months, and for not-so-efficient organizations, it can go up to 18–24 months. It will entail working across multiple workstreams like data quality improvement, data governance, creation of enterprise-wide data architecture and metadata, and having the right infrastructure and IT infrastructure for supporting these.

A big chunk of data leads has organically grown from existing IT and data organizations of banks. Our experience suggests that a significant portion of these leads have not gone through enough training to think at the strategic level. This also means data organization is handled in piecemeals and becomes an ever-evolving exercise. While data management will always be ever-evolving, the difference between mature and immature organizations can be gauged by whether the milestone of enterprise-wide data access of good quality and in a secure environment is available to users. The above statement has subjectivity, but this is a big differentiator for organizations, so organizations should rather spend on getting a lead, who has seen it for a world-class organization and has a sense of how it can take their organization closer to the world-class organization.

4.10 Evolution of Best-in-Class Data Organization

Best-in-class data organization has a good blend of business and data team collaboration. The data organization will have a few dimensions:

Governance

There exists overarching data governance for the FI. The compliance function is one of the areas where it is commonly applied. All data journeys are deduplicated to a major extent. i.e., risk, finance, fraud, and AML, all share a big chunk of the data commonly. Few specifics for each domain are managed through a flexible yet enhanced metadata management layer and eventually data cube layer. There exists clear data governance, data protection policy, and data stewards who ensure management of data lifecycle and take responsibility for securing it.

Infrastructure

We have been hearing that storage capacity costs have come down significantly in the last few years. Storage is a commodity. A good quality data organization will have a sound capacity to host structured and unstructured data for the business. Similarly, the sizing of processing is done for regular and irregular activities with multiple user access. If the data is hosted on the cloud, FI doesn't need to invest in infrastructure for short-term projects requiring large data storage and processing capability. FI can leverage Kubernetes type of structure for the scale-up and scale-down for such needs.

People

A very important aspect of such an organization is the leader and its team. A leader has a good appreciation of the support it needs to provide to the business and has a decent domain understanding to understand and drive the organization. The team also has resident business analysts who understand the domain and data. Lastly, there exist good quality data engineers who can work on the data pipelines and leverage it for productionizing the solution on the scale.

Teams are regularly upgrading themselves with recent changes in technology and business.

Data Architecture

Data for all compliance functions is managed centrally. Data catalogs, data quality rules, and derived variable definitions are transparent and are available to all teams in an easy manner. All data is desiloed and operates on the dev-ops model. Roles, responsibilities, and processes for data enhancement, data corrections, data updates, data purge are clear and well-documented.

There exists clear layer of data for analytics and machine learning perspective and clear data architecture on production, which can be leveraged for migrating any insight into production with minimal effort.

Planning for AI in Financial Crimes

5

5.1 Introduction

Artificial intelligence in financial crimes has become a topic of interest for all professionals across financial institutions. Regulators are now expanding their ambit on every aspect of financial crime. To give context to the need for AI in financial crimes, let us understand the evolving landscape for financial institutions.

5.2 Forces Shaping the FCC Ecosystem

There are three major areas that are shaping the ecosystem—(a) the evolution of the banking and payments industry. These have resulted in increased products, channels, and processing partners. This has hindered the visibility that a financial institution gets on the customer and its transactions (b) Increasing financial integration across markets and a rising appetite for transactions by individuals and corporates. The world is more connected. A few have a genuine need to do business even with high-risk countries and entities, thereby making the task of financial institutions tougher and (c) regulatory oversight—there are multiple instances of central bank's audit team of 40 + auditors reviewing a large bank with compliance in financial crimes as the only theme. Federal investigations extending to links in financial institutions are a common theme. This has necessitated FIs to be more vigilant and cautious while processing millions of transactions and customers every day.

This is further complicated by areas that now warrant investigation. Starting from normal cash, wire transactions to correspondent banking, digital wallets, payment gateways, wallet, and prepaid mobile top-ups, and white goods financing; the ambit of payment and banking solutions are many. Central banks are now prescriptive on coverage domains that any compliance department of financial institutions needs to manage (Fig. 5.1).

A. Gupta et al., *Artificial Intelligence Applications in Banking and Financial Services*, Future of Business and Finance, https://doi.org/10.1007/978-981-99-2571-1_5

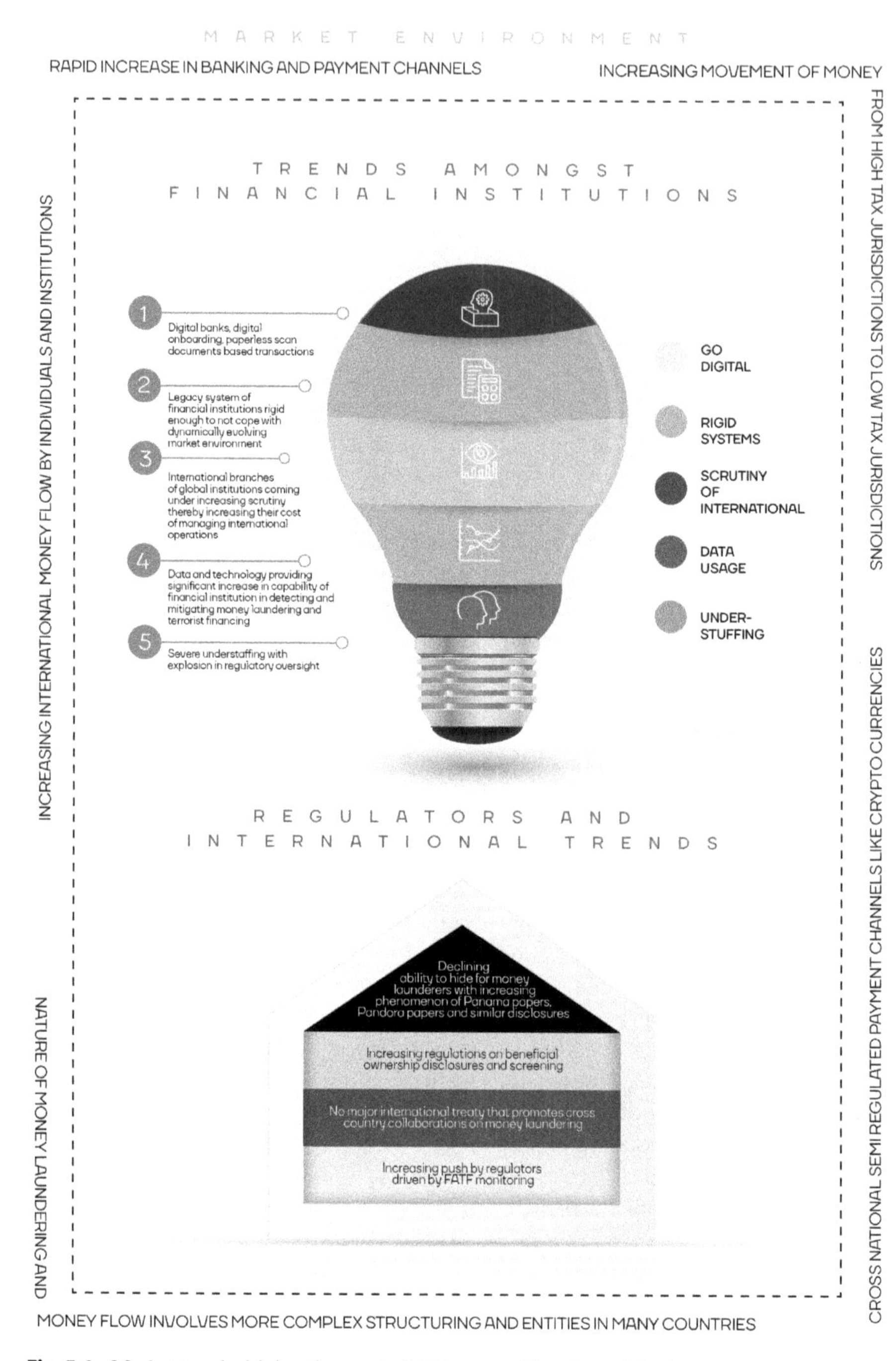

Fig. 5.1 Market trends driving the need of AI in controlling financial crimes

When the foreign account tax compliance act (FATCA) and common reporting standards (CRS) came into being, the efforts for compliance seemed like a big overload. In hindsight, they now seem tiny as compared to compliance to AML and combatting of financing of terrorism (CFT).

Most of the heads of compliances that we spoke to were clear that the workload with new guidelines is not feasible.

It is not the case that regulators are happy either. As per one of the recent studies by Deloitte, a big chunk of SAR reporting to the law enforcement agencies are of little value for them. It is primarily due to the quality of information filed in the SAR filing. Organizations being conservative also have an implication on false positive rates of SAR filings. While the number of false positives within SAR filings varies by jurisdiction, it is undisputed fact that the majority of the SAR filings eventually turn up to be false positives.

Keeping all the above challenges in mind, the regulators, as well as financial institutions, are forced to look at AI as the budget for their departments is bloating and this is still increasing.

However, you would have noticed in the previous chapter that there is a resistance to moving the route of digitization and bringing more robotics and intelligence into their work.

5.3 Pitfalls to Avoid When Designing AI-Enabled FCC Organization

All compliance officers have either invested or are in the process of investing in new-age solutions. However, with few failures or the inability to realize expected or promised benefits, they are also wary of moving at a fast pace. Here are a few important tips for the chief compliance officers, who intend to leverage AI in their business:

- Every organization, depending on its business has its own opportunities and constraints for leveraging AI successfully. Do not copy others. Assess the opportunities in your organization and the state of enablers and people. Based on an objective assessment of the current state and opportunities, develop a road map and also define milestones.
- Define a few operational and financial metrics to measure outcomes. Investment into these capabilities is meaningless unless there are tangible outcomes
- Have a clear view of roadblocks, state of data, level of integration, the capability of teams, turnover of teams, and so on. Work with a clear focus on outcome-based plans
- Create risk-reward for the managers who are overseeing digitization and AI initiatives. We have often seen that managers are working part-time on these initiatives. This doesn't help provide push and outcomes for these initiatives
- Be realistic about the outcomes. We have witnessed two extreme kinds of management styles that need correction—one that is partly hands-off after defining

the expectations. Business managers are part-time overseeing the initiatives. Initiatives are generally delayed, have quality issues, and are not accepted by the rest of the organization. Eventually, investment does not deliver value. Another management style is very aggressive. We have heard this statement from many CCOs—"we wanted to go live with this initiative in two months. We also want a 30% reduction in false positives. Make it happen" This is another recipe for failure. Have realistic estimations based on the situation on the ground and expectations of the project. Do keep in mind that the risk of misclassification is quite high. Focus on test and learn, be conservative, and get confident about the intelligence and the outcome, before full-scale rollouts

- Finally, AI projects need to be jointly led by business and IT. An interesting situation in some parts of Asia is that the AI projects are led by IT. While from enterprise architecture, technical requirements, data, and information security perspective, IT and data teams are key stakeholders, they are not the owners of this type of project. The outcome of such governance is that the head of the project or main sponsor driving such a project does not have a vision for the organization. Modular approach rather than digital transformation after proof-of-concept results in a lot of organizations ending up with mediocrity. An optimal governance structure is the business teams, data, IT and AI team members jointly form the core team for such governance.

5.4 Setting up Roadmap for AI Organization

Steps for developing an AI roadmap for the compliance function.

1. Defining digital transformation end state
2. Opportunity identification
3. Use case conversion
4. Readiness assessment
5. Prioritize initiatives
6. Roadmap.

The first stage in this type of setup is defining the end state of a digital transformation. What will digitization mean in terms of process, technology, and its impact on people? Which part of the process can go through automation, which activities can get more intelligent? What role will technology play and what will be the role of people in the whole ecosystem? How will people leverage technology, and a sense of potential benefits would be the key elements of the digital transformation? A blueprint of this end state will provide clarity to the organization in terms of the direction they want to pursue. It might be the case that the end state is almost similar across organizations of similar type. However, the prioritization, the journey, and the level of technology leverage can change across an organization. Hence, it cannot be a cookie-cutter same size fits all approach.

The next stage is opportunity identification. Once the end state is defined, based on the size of the opportunity, process, regulatory constraints, and organization's culture, it is imperative to identify digitization opportunities as well as AI opportunities (Fig. 5.2).

The opportunities need to be converted into use cases to ensure that there exist tangible outcomes and benefit quantification. The use case cannot be generic. It must clearly define inputs for example data, technology platform, infrastructure, process, and system impact. As consultants, one of the major challenges faced is the realization of dependencies in the middle of the use case implementation. Consultants are hired for a specific use case. The client initially states that all inputs and integration are ready. Then in the middle of the project, the client is hit with the realization that the prerequisites are not fulfilled. It results in unnecessary delays, and unnecessary pressure on teams from senior stakeholders, as junior team members do not admit the lack of preparation in front of seniors. If the building blocks are assessed upfront, the use case implementation moves smoothly

Prioritize initiatives—based on impact, readiness, and time to execute, the team needs to prioritize initiatives.

Roadmap—based on the improvement of infrastructure, data quality, and capabilities, the teams should create a roadmap to traverse the journey to achieve best-in-class organization for compliance. When the team starts with the end state in mind, the tangible milestones are easy to manage and define in terms of targets and budgeting.

Compliance function in a financial institution is heavily scrutinized. The risk appetite of stakeholders is very low. This needs to be factored into the journey, as traditional digital transformation timelines cannot be applied. While the transformation will follow an incremental approach, the adoption of digitization must

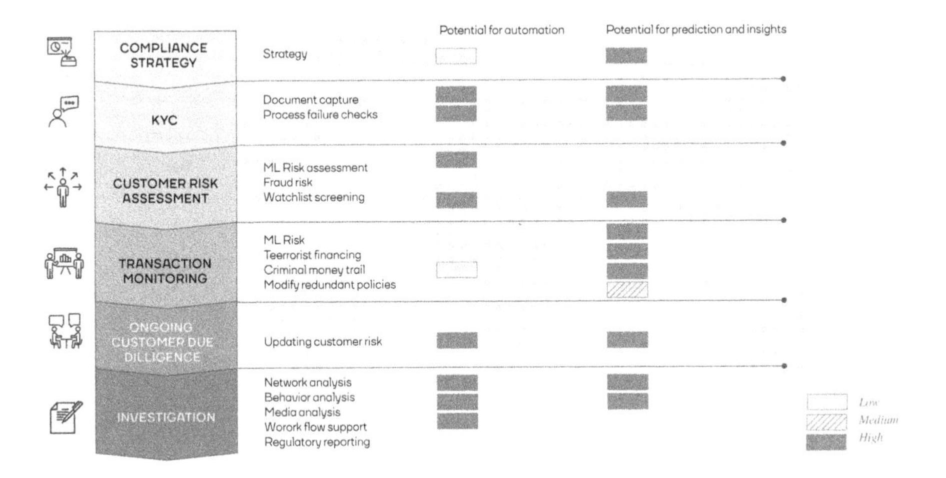

Fig. 5.2 Showing the digitization opportunities in the customer lifecycle

be slow. Neither the bank's team nor the regulators are comfortable with a fast-paced transformation. Concepts like minimum value product (MVP) may not fly that easily in the compliance department. Nevertheless, slow and steady will win the race.

5.5 Building Blocks of a Best-in-Class AI-Enabled Compliance Function

Artificial intelligence-driven organizations typically reside on four major pillars:

Technology—a digitization journey cannot move without the right technology. Authors, keeping in mind the importance of this have dedicated a chapter on technology solutions. The only item to add here is one important consideration. All organizations have legacy systems. Those cannot be done away with. The roadmap created should have a clear path for technology. While new technology will always keep coming and replacing legacy platforms, the journeys are often long and bear a significant risk of delivery failure. Keeping these in mind, CTOs and business sponsors should keep the continuity in mind. One good option could be plug-and-play integrations. Fintechs and Regtechs bridge these gaps very effectively. An example could be an alert optimization solution. Machine learning or analytical model-driven solutions cannot replace the existing solutions. Challenge is that existing solutions are rigid, and they cannot accommodate these smarter solutions. This is where a smart architect is required, who looks at integrations through "plug-and-play" mode, rather than always starting with defining the need for replacing the solution to make it work (Fig. 5.3).

Policies and processes—as mentioned in the section above, the transformation journeys have a critical need for mapping the right policies and processes. Artificial intelligence can put a mechanism to measure breaches of controls on policies and help identify opportunities and redesign the digitized process. Elements of benefits accrued to the business are defined upfront. It ensures that the stakeholders can view the end state and the benefits it will bring.

Data—data is another important foundation of a successful AI platform. Given the importance of the data, the authors have dedicated a chapter on data readiness.

People—people are one of the most important foundations of any organization. AI or digitization department is no different. The right skill set and the right team can make the initiative successful even if other elements are not ready. A wrong team can spoil millions of dollars and thousands of hours of investment. Given the importance of people and organization design, the authors have dedicated a chapter on people and organization.

TECHNOLOGY

- Enabling right infrastructure
- Enabling the right technology platform – segregated into machine learning, model ops, visualization and BI tools

DATA

- Enabling good quality, single view of customers, transactions, networks
- Enforcing strong data lineage, cataloguing and centralized availability for data access
- Data models enabled for minimal rework

POLICIES AND PROCESS

- Developing sound redesign of process with intelligent solution
- Capture manual intervention to assess improvement areas
- Develop second and third layer of controls into the process

PEOPLE

- Enabling team members with the relevant skillset
- Enhancing skillsets for the teams by hiring relevant skills not available in-house

Fig. 5.3 Organizational improvement required to maximize impact of AI in an organization

Applying Machine Learning for Effective Customer Risk Assessment

6

6.1 Introduction

A compliance journey for any customer in a financial institution starts with customer onboarding. Regulators expect financial institutions to collect the information of their customers and update it on a regular basis. They are also expected to assess if the customer is worth dealing with or if the customer is expected to indulge in some wrongdoing or possibly bring disrepute to the organization by involving itself in some illegal activity. While a financial institution would not want to deal with such customers, sometimes lack of judgment or greed of front-line staff could result in acceptance of such customers. This is where the regulator had to push FIs to follow a structured approach and follow the prescription of the central bank.

In almost all jurisdictions, regulators have instructed banks to collect information on the customer through a process called know your customer (KYC). Financial institutions are expected to collect the identity proof, source of income, purpose, details of transaction types with the institutions, tax details, and disclosures on tax residencies in the USA and other countries. It enables the financial institutions and eventually the regulator to compile information about various financial relationships that any individual has in the country and in a few cases across countries (FATCA or CRS).

FIs are also required to assess the customer risk from a financial crime perspective as well as confirm that the customer is not a sanctioned or watchlist entity.

There is another important element within the customer risk assessment, which is called ongoing due diligence or dynamic customer risk assessment. This activity entails reviewing and updating the customer risk score for committing financial crimes, based on the customer's behavior with the financial institution.

A. Gupta et al., *Artificial Intelligence Applications in Banking and Financial Services*, Future of Business and Finance, https://doi.org/10.1007/978-981-99-2571-1_6

While these guidelines are well-intentioned, each of these activities is cost-intensive for the financial institution and impacts the customer experience. Some of the major pain points for this activity are:

- The manual process of data and document collection and verification. The KYC documents are re-updating KYC is generally a manual process. Either they are scanned by the FI staff, or they are scanned and sent by the customer through an attachment to a mail. The manual process entailing an overview of such documents and processing them is huge
- Lack of clarity and clear policy on aspects to look for in customer risk assessment. There are instances where the compliance officer keeps asking for documents and supporting documents and clarifications from the customer. Sometimes they are pointing to investigations, in other cases, they are probing and seeking documentation to be sure. Irrespective, the lack of clarity in terms of the process for enhanced due diligence is painfully long and unstructured. There exists a huge opportunity to optimize this for the FIs.

Imagine the above challenges coupled with the volume of customers for whom the organization needs to conduct these assessments. It consumes almost 25–40% of the total workload of the compliance function. Let us understand each of the activities conducted by the compliance function. The authors will then delve into the application of digitization and artificial intelligence in each of these activities.

6.2 Know Your Customer (KYC) Processing

As mentioned, a financial institution is expected to conduct KYC for all the customers onboarded. FIs are also expected to capture updated KYC for customers, depending on their onboarding or ongoing risk on an ongoing basis. Depending on the jurisdiction, it can either be for all the customers on a regular interval or vary based on the captured riskiness.

FIs are also subjected to internal audit reviews on ensuring that there are minimal breaches in conducting or updating KYC as it can result in severe audit comments by the central bank or the regulator. Till a few years back, this process was manual. This has resulted in huge manpower costs for conducting the exercise. There were significant oversights pertaining to the timeliness of conducting those KYCs. There are high instances of company registration renewals, Government ID information, and similar information not updated, post the expiry of such documents. There are also several instances of bank officials and customers doing multiple back and forth for conducting such activities.

Technology has come to the rescue of compliance professionals. Starting from the launch of e-KYC where the information is seamlessly collected from different IDs available to the financial institutions through direct integration with government portals. It means that once a customer has authorized the financial institution to collect the information on his/her behalf, the financial institution, automatically pulls and fills the requisite information in their forms. Documentation requirements are reduced as the information is digitally collected from a reliable source.

6.2.1 Automated Alerts for Expiry and Renewals

Based on the information updates, systems can be entrusted to generate auto alerts for the relevant staff and the customer that the documents are up for renewals, and they can be sent.

6.2.2 Automation of Information Extraction

For a streamlined process, another area of opportunity is the automation of information extraction. This can be achieved through scanning the e-mails, sorting, triaging, and extracting the relevant information. Human intervention can be in terms of quality control on such information collection. This technology application can save millions of hours across FIs. The important aspect of this type of solution is the accuracy of algorithms, as poorly written algorithms can spoil the trust of such systems and can also significantly increase human interventions in such processes.

6.2.3 Computer Vision Application on e-KYC

A further improvement from AI in this process is the application of computer vision in the process. FIs are now able to investigate the consistency of uploaded documents, e.g., whether the National ID mentioned in the form by the frontline staff is the same as the one in the document uploaded, etc., through computer vision. Discrepancies can be alerted and then investigated. Another benefit of using AI in this area is that it provides quality assurance on 100% of the transactions.

Adopting technology and sophisticated algorithms can take away big pain, cost, and customer inconvenience in these processes.

6.3 Sanctions and Watchlist Monitoring

It is the responsibility of the financial institution to ensure that the customer they are onboarding, does not belong to any list of watchlists or belong to a banned entity. In this regard, there are a lot of agencies like EU that prescribe its watchlist. There are around ten similar entities like FBI, OFAC, UN watchlist, UN Vessels watchlist, and so on. Alongside, most of the regulators also have local watchlists.

A financial institution needs to ensure that any customer engaging with it, is matched against these lists. If there is a credible match found, the investigation team needs to ensure that the matched customer is not genuinely on the watch list. The investigation team must investigate each of these cases manually. This is a time-consuming process. There is a similar real-time matching exercise that happens during every remittance transaction as well. This is one of the most challenging areas for a financial institution. A big chunk of workload for the investigation team arises from this area.

It is easier than beneficiary match during transaction monitoring (we will discuss the challenges later when we discuss transaction monitoring in the next chapter), as the customer provides a lot more information to the financial institution like name, tax id, identity proof, nationality, and original birthplace. This means that the matches have a lower false-positive rate.

A lot of software solutions focus on improving the efficiency of the matches. A typical phonetics or linguistics match or the simple combinations (ensembles) result in almost 60–65% false positives for a 100% recall (no miss in the actual matches). As we mentioned, it is rather easier for banks to match the applicant's name against sanctioned entity given the information availability being higher (additional information like tax id and National ID); however, for financial institutions like money exchanges, even these fields are not sufficiently available. Keeping this in mind, most financial institutions end up matching names or applicants or entities to generate alerts.

6.4 Expectations from Compliance

If one is a compliance team member, one doesn't necessarily need to understand all the theoretical construct that is explained below in this chapter for name screening. This would be of interest to data scientists who are either evaluating or developing a proprietary system for their financial institution. It is a good read for you irrespective. One piece of information which is critical for you is still the metrics—the false positive rate for a 100% recall name matching (100% recall means that there is no miss on an actual match that should have happened). Benchmark this and let any incumbent challenge your existing solution. Till they can beat the benchmark false positive rate, it would mean lesser work for your teams and cost savings.

6.4.1 Challenges for Name Screening

Typical challenges in name matching are given below:

In the case of organizations, there are a few additional challenges that also need to be handled. Some of the typical ones are listed below:

Sharing the same name with a different incorporation type. This is typically observed in international databases. A company name is registered with the company registrar in a particular country. However, the same name can be registered in another jurisdiction without modification. This can result in false matches. Another reason for these false matches can be a company having subsidiaries or branch offices in different countries. Since the name, in this case, could be common, hence it can result in false matches (Fig. 6.1).

Sharing common names—for example, "ministry of", "high street", "steel manufacturing", country name LLC, etc.

6.4.2 Approaches for Name Screening

Figure 6.2 provides the details of typical matching algorithms prevailing in the market:

The most popular mechanisms available in the market are:

Phonetics-based—these algorithms focus on "sounds like" functions to match similar-sounding names. Typical algorithms that are popular in phonetics are:

Edit distance based—these classes of algorithms focus on a number of edits that need to be made in any name to match the other name. For example, if we have to match John to Jenson, we will need to make five edits to match it. As it is apparent, the higher the number of edits, the lower the chances of a match.

n-gram algorithms—these classes of algorithms convert the name into a group of strings of alphabets. Then the groups are matched, individual. Let us take the same example as above—John matched with Jenson. If we create 2-g matches, it will mean John having Jo, oh, hn as three combinations of 2-g. Jenson will have Je, en, ns, so, and on. Now we can calculate the match percent. If you would notice, Jo, oh, and hn are not found in the 2-g of Jenson and hence the match is 0%.

Then there are modified algorithms leveraging similar logic for matching along with minor modifications to take care of issues that the proponents of those algorithms have found.

Rule-based matching

This is one of the most rudimentary forms of matching algorithms. It tries to maximize the true matches by looking at the name's patterns. As you would imagine, while results from this type of algorithm will be very good, the challenge would be that it would be considered as an overfitted algorithm. It would be overfitted for a simple reason, the algorithm works well for given patterns of names and issues with spellings. If the patterns change, the name matching will deteriorate very fast.

Ensembles

As it is apparent from multiple algorithms above, that none of the algorithms work well on different name combinations. Different ethnicities have different challenges from name-matching perspective. To improve the matches, the data scientists started evolving the name matching with something called fuzzy matches. These fuzzy matches leverage combination of different algorithms into ensemble.

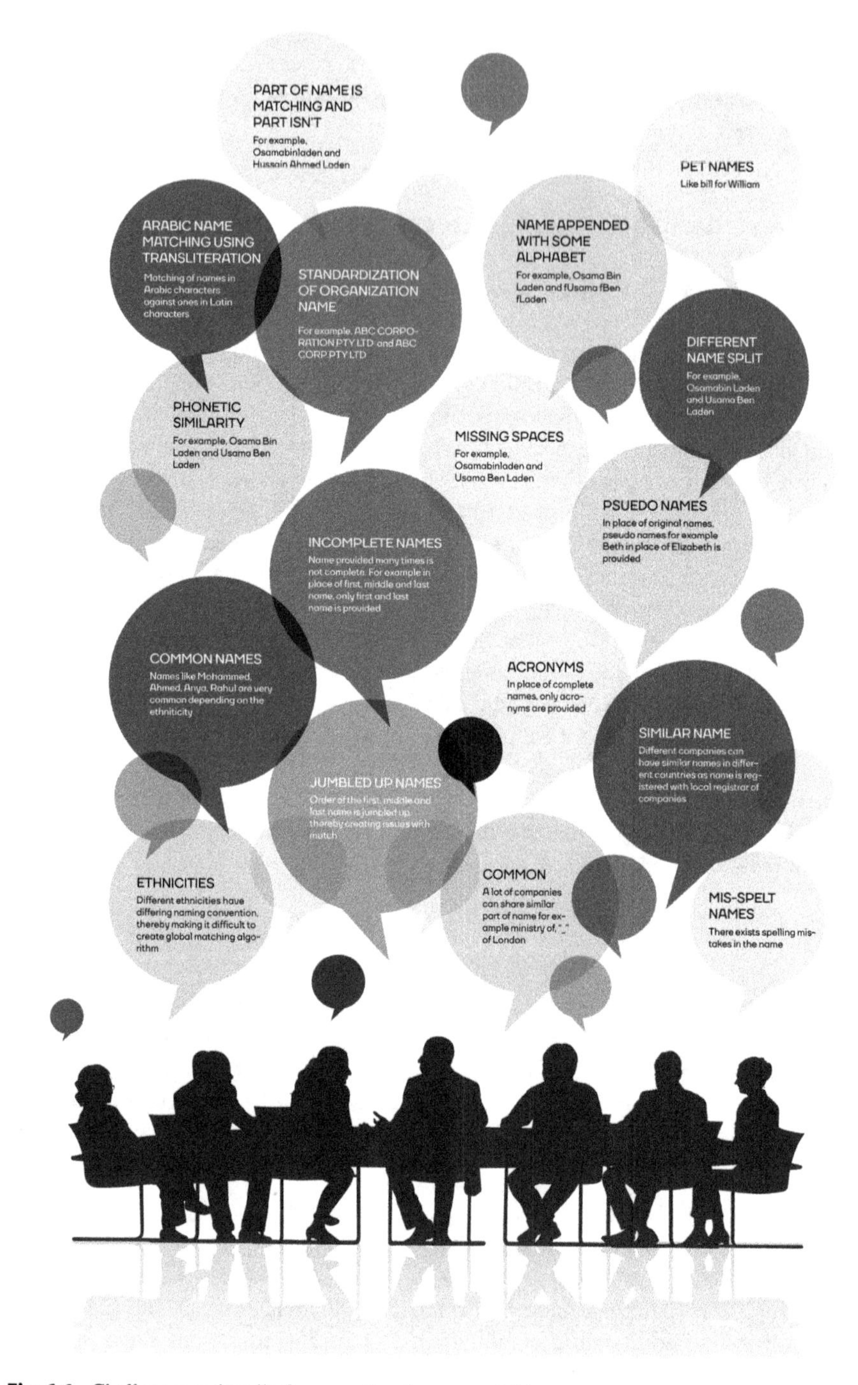

Fig. 6.1 Challenges to handle for an optimal name matching

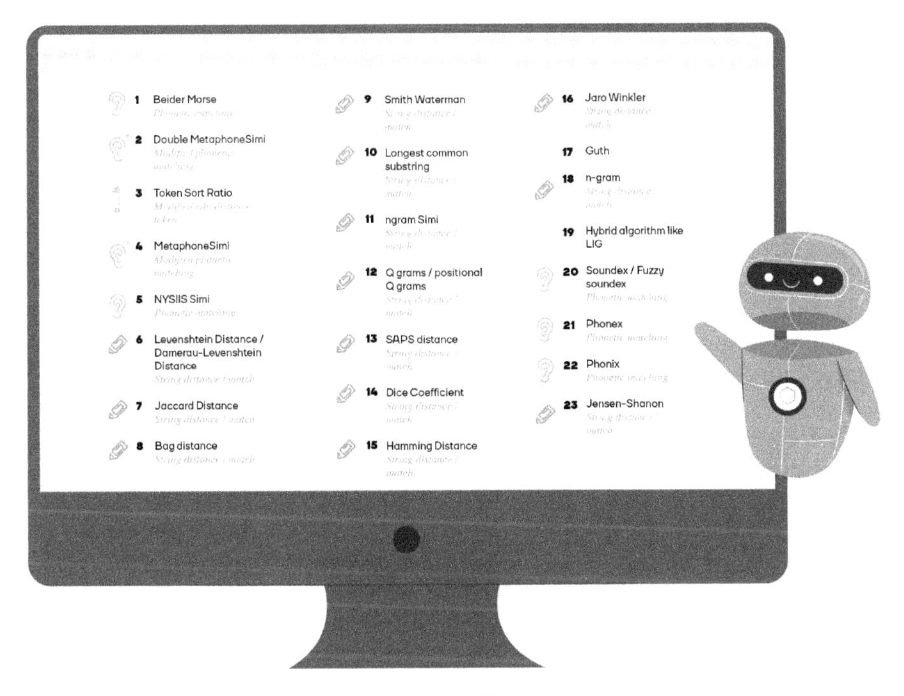

Fig. 6.2 Popular name-matching algorithms available

So for example Levenstein algorithm will be combined with n-gram and the match will be given a score. Practitioners also try to define some business rules depending on the input names.

A lot of practitioners figured out that typical misspelled names have wrong vowels. To avoid mistakes, they also eliminate vowels and then match the names.

Professionals will try to hardcode a few rules for example, in the Middle Eastern countries, the name "Mohammed" is written differently. Like Muhammad, Mohamad, Mohammad, Md. Another challenge is with the name Ahmed, which again has multiple similar spellings and has a close lineage to Mohammed. We have seen few institutions creating master databases to lookup standardized values. This is also applicable to the commonly used short form of names, for example, William becomes Bill. In those cases, name matching won't be able to generate a match, unless there is a lexicon of such names. It happens with company names also.

All the above implies that individual name matching and entity name matching are slightly different variants. A data science team can profile names and can improve even the best-performing algorithm that best suits their portfolio.

Machine learning algorithms

ML takes this one step further. Unlike the ensembles, machine learning models can further improve the matches. There are multiple ways in which the matching algorithms can be improved.

Improvisation over ensembles—in this case, a different combination of algorithms can be developed using machine learning to assign weights or importance to each algorithm. ML, in this case, learns to reduce false positives, by assigning different weights (as against the weights or combinations defined judgmentally).

Feature extraction and then model development—in this case, the features of the names are extracted and then used to train the model for name matching, for example, length of names, number of characters in each name, match to any incorporation type, etc. A user can take such a list as an input and then develop a name-matching algorithm. The objective of machine learning, in this case, will also be to reduce the false positive.

6.4.3 Impact of Machine Learning Algorithms on Name Matching

ML has improved name matching significantly. Our team has worked on a patented algorithm that has improved the false positive rate from around 40% to around 20% (for 100% recall) using machine learning algorithms. 100% recall means, that we are not leaving any true positive match that should have been matched. ML models will always be like a recipe available to a cook. There is no rigid rule in terms of what will provide the best results. However, the ability of the modeler to understand features that can be applied across ethnicities, learning from false positives, and further developing an iterative process for the model to continue learning are key to success.

6.4.4 Process for Training Machine Learning Algorithms for Name Matching

Figure 6.3 shares the mechanism for the development of a machine learning algorithm:

The first step in this would be transliteration. Names in watchlists appear in many languages. While a language-agnostic name-matching algorithm can be developed, that would essentially mean that the lexicon on which the name-matching algorithm should be trained has to be huge. Another challenge with watchlists is that depending on the regulator, the watchlist can have an English name or name in a foreign language. For example, it is possible that Osama Bin Laden is available in both English and Arabic names on the watchlist.

To handle such issues, it first needs to be transliterated to English. For every language, there exist multiple works of literature that deal with this topic. As you would imagine, transliteration would require going through the literature of

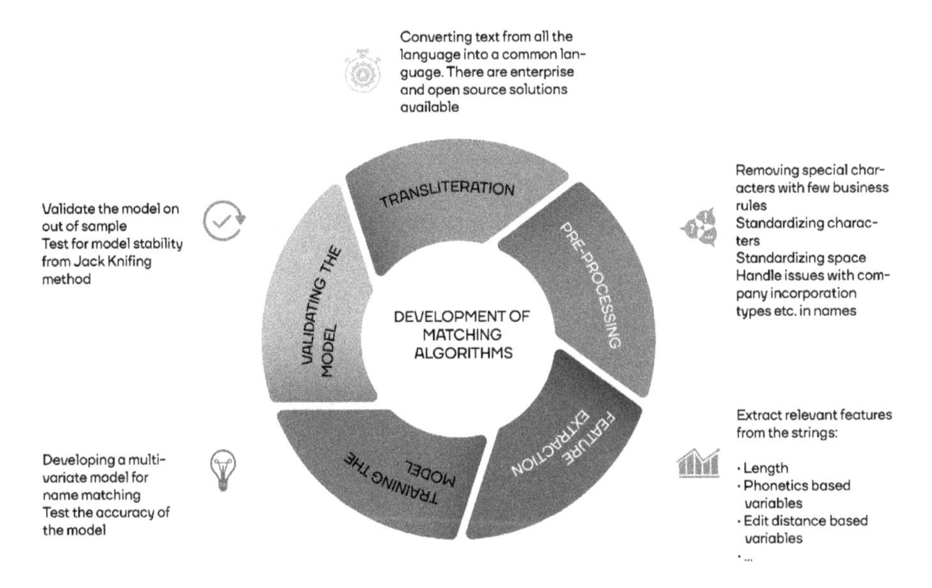

Fig. 6.3 Steps for developing a machine learning model for name matching

every language as one needs to understand the script, specificity of language, and pronunciation of each alphabet. Thankfully for us, a lot of open sources and enterprise-level solutions handle transliteration.

The next step in a model development exercise is the preprocessing of the names. Names or phrases will have elements like special characters, hyphens, extra spacing, salutation (Mr, Ms, Dr.), and so on. First, these attributes mentioned should be eliminated to ensure that the names to be matched are not infested with such problems and are more standardized. It can be handled through the elimination of special characters (even here, there are a few linguistic rules for example "—apostrophe", and "—hyphen" sometimes have a meaning). One cannot simply eliminate them. Hence, there exist few business rules that do the preprocessing of the text. Standardization entails bringing all names into common character. If there is an acronym, then these are marked.

Splitting the sample into development and validation samples before a model training process is started.

The next step is feature extraction—depending on what a modeler wants to train its model on, the next step would be to extract features. As we discussed earlier, some of the common features are—the length of the name, the number of characters in the name, one can possibly mark ethnicity based on a few business rules, whether it is a company or an individual, *n*-gram, scores of algorithms, and so on. Once these are extracted, the next stage would be model training.

Model training essentially means that the features extracted are used as independent variables. The outcome is a true match or a false match, which is what a machine learning algorithm is trying to optimize. All possible machine learning techniques like logistic regression, decision trees, random forest, XG Boost, and neural networks can be used for training the model.

The outcome of the model will be assessed on various modeling parameters. However, precision and recall are the typical metrics used for the assessment. Precision refers to the percent of the outcome correctly classified as a percent of the total classification of the model. Whereas recall is the percent of total true positives classified. For example, if there are supposed to be 100 true positives in a table for name matching. The model predicts 70. 50 of them are actually true positives and 20 are false positives. In this case, the precision of the model is 50/70 = 71.4%. While only 50 out of 100 true positives were detected, that makes the recall to be 50/100 = 50%.

Along with the precision and recall which are very important business aspects of the model, a modeler also needs to assess the statistical significance of the variables or features of the model. They need to be statistically significant for at least a 95% confidence level. Lastly, a predictive power measured by GINI or Kolmogorov–Smirnov (KS Statistic) can also be calculated. While in the above-mentioned type of models, predictive power is of slightly lesser consequence. Precision and recall will be of utmost value, but it is still good to calculate for benchmarking purposes.

Validation of the model is the last step in a model development exercise. Once the results of the model are inferred and the modeler is convinced about the quality of the model, the model is then validated against the out-of-sample validation. What any modeler will look for is the change in the precision for a given recall in the model. A well-trained model will have a stable precision for a given recall in an out-of-sample validation exercise. An overfitted model will witness a drop in the precision of the model for a given recall. This can also be seen from drop-in the other prediction metrics like GINI or KS Statistic.

A better-performing model from a prediction perspective results in lower false positives and hence a lower number of investigations. Assuming around 4–5 min of average investigation time (few might take less than a minute, while few others could take up to 30 min), if a machine learning algorithm can reduce ~ 200,000 false matches without compromising on true positives, it could mean up to 10–12 FTE less requirement for investigating those false alerts. Depending on the markets, the cost can vary. But for medium- to high-income countries, this will translate into approximately half a million dollars. The magnitude of this investigation increases manifold in counterparty transaction monitoring. We will discuss this in more detail in a later chapter. But one can imagine, the load on financial institutions that could be potentially reduced by applying machine learning algorithms here.

6.5 Customer Due Diligence

Customer due diligence (CDD) during onboarding is one of the most critical areas for an FCC organization. This remains true despite significant investments made in upgrading the technology platform and sound compliance during onboarding and ongoing due diligence. Let us start by mentioning the core problem of such mechanisms. The objective of such an exercise is to mark the high-risk customers for higher surveillance on their financial transactions. However, during data analysis, we have conducted rank ordering of assessed riskiness of customers to their actual SAR filings or situation when the case is created in case management for them.

Our analysis revealed that there does not exist a high correlation of riskiness of the customer assessed by FIs to their actual risk behavior. This puts in question the whole risk assessment process for the majority of the FIs.

Before going deeper into our proposition for such financial institutions, let us first understand the existing process followed by various financial institutions:

The manual nonstandard process of customer risk scoring—Believe it or not, this is still a prevalent mechanism for a lot of banks. There exists a loosely defined dimension and definition of risk on a few dimensions like industry, products, jurisdictions, and so on. A compliance officer asks for information and assesses a customer judgmentally on these dimensions. If the compliance officer perceives it to be of higher risk (because of industry or jurisdiction or a combination thereof), the officer initiates enhanced due diligence.

Problem is that the officer is now trying to investigate every possible information of the individual/organization and their related parties. Requests can vary from tax filings to bank statements, to providing an explanation of every line item of a bank statement, detailed biography, and many more. The compliance officer is trying to look for something suspicious in these instances and satisfy himself with the genuineness of the customer. Irrespective, the process is time-consuming and painful for the customer, as there is no prescribed list of what the compliance officer is supposed to be looking for. It is all a function of his/her experience and anecdotes of what could result in finding something suspicious. The outcome of such an exercise is a judgmental risk grading. If one asks two compliance officers to compare the rating and justify the given rating, rationales and outcomes could be highly subjective.

CDD judgmental scorecards-driven processing—This is now the logical next wave, that is being pushed by a lot of regulators. Regulators have realized the futility or the lengthy process of such an assessment. They are pushing the financial institutions to standardize their KYC process and ensure relevant information asked has a meaning in their risk assessment efforts. Information captured is fed into the scorecard. CDD scorecards are divided into two types:

1. Onboarding CDD scorecard
2. Ongoing due diligence and dynamic customer risk rating.

6.5.1 Onboarding CDD Scorecard

The scorecard strives to provide a risk rating to every customer from a financial crime perspective. It relies on various factors like demographic, banking, product-related information, whether politically exposed PEP, nature, and quantum of income, and so on. Depending on the sophistication and nature of FIs, these can either be statistical scorecards (machine learning-driven) or they can also be expert-based scorecards. Expert-based scorecards are still an improvement over rudimentary nonstandard processing of customer risk assessment. It has the advantage of exactly knowing what to ask from the customer, the way the information will be processed and fed into the scorecard, and finally the scoring outcome. Another advantage is the reduction of subjectivity. Two compliance officers can provide input on the same customer and can get the same risk rating. It can also be automated to a major extent or completely, the risk assessment process for a big chunk of onboarded customers. Depending on the goodness of the scorecard, medium, and high-risk customers can go through manual process-driven due diligence.

One important part of this exercise would be to streamline areas to watch out for and infer from the facts that can further reduce the subjectivity and randomness of questions asked. A compliance officer should be entrusted with going out regular questions, requests for documentation, and more investigation, however, organizations shall clearly define the governance and control around it (Fig. 6.4).

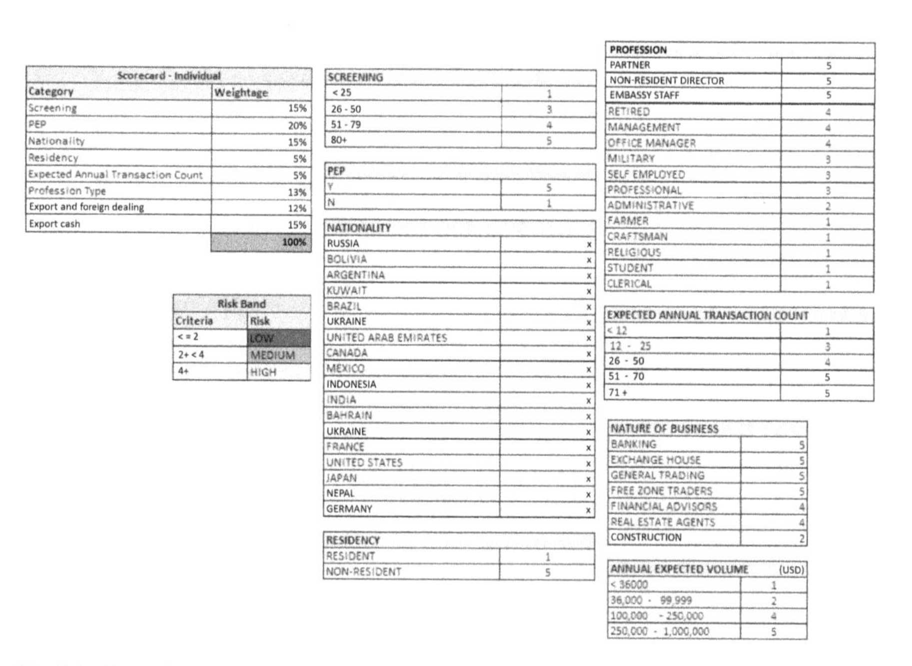

Scorecard - Individual	
Category	Weightage
Screening	15%
PEP	20%
Nationality	15%
Residency	5%
Expected Annual Transaction Count	5%
Profession Type	13%
Export and foreign dealing	12%
Export cash	15%
	100%

Risk Band	
Criteria	Risk
< = 2	LOW
2+ < 4	MEDIUM
4+	HIGH

SCREENING	
< 25	1
26 - 50	3
51 - 79	4
80+	5

PEP	
Y	5
N	1

NATIONALITY	
RUSSIA	x
BOLIVIA	x
ARGENTINA	x
KUWAIT	x
BRAZIL	x
UKRAINE	x
UNITED ARAB EMIRATES	x
CANADA	x
MEXICO	x
INDONESIA	x
INDIA	x
BAHRAIN	x
UKRAINE	x
FRANCE	x
UNITED STATES	x
JAPAN	x
NEPAL	x
GERMANY	x

RESIDENCY	
RESIDENT	1
NON-RESIDENT	5

PROFESSION	
PARTNER	5
NON-RESIDENT DIRECTOR	5
EMBASSY STAFF	5
RETIRED	4
MANAGEMENT	4
OFFICE MANAGER	4
MILITARY	3
SELF EMPLOYED	3
PROFESSIONAL	3
ADMINISTRATIVE	2
FARMER	1
CRAFTSMAN	1
RELIGIOUS	1
STUDENT	1
CLERICAL	1

EXPECTED ANNUAL TRANSACTION COUNT	
< 12	1
12 - 25	3
26 - 50	4
51 - 70	5
71 +	5

NATURE OF BUSINESS	
BANKING	5
EXCHANGE HOUSE	5
GENERAL TRADING	5
FREE ZONE TRADERS	5
FINANCIAL ADVISORS	4
REAL ESTATE AGENTS	4
CONSTRUCTION	2

ANNUAL EXPECTED VOLUME	(USD)
< 36000	1
36,000 - 99,999	2
100,000 - 250,000	4
250,000 - 1,000,000	5

Fig. 6.4 Example of the expert scorecard—sanitized example

6.5.2 Machine Learning Models for Customer Due Diligence

The difference between a machine learning model and an expert scorecard primarily resides with the process followed and hence the power of prediction. An expert scorecard relies on the collective judgment of various compliance officers. There exists little or in a few cases, no data-driven insights for such a classification. Whereas a machine learning model relies on data-driven insights and generally has better predictability of financial crime perpetrators.

Machine learning models typically go through a standard model development lifecycle. For the sake of avoiding duplication, readers can refer to Chap. 8 where the model development lifecycle is explained in detail.

6.5.3 Ongoing Due Diligence and Dynamic Customer Risk Assessment

One can relate to this aspect of customer risk assessment from the level of information available with any FI about its customer. FI has very little behavioral information about any onboarded customer. Even if a bank statement is asked, it is never clear that the statement pertains to the primary bank, where a customer is transacting. Keeping such challenges in mind, the risk assessment capability of the FI on the customer is limited at the time of onboarding. However, this changes when the customer starts transacting with the FI.

The turnover, mode of payments, parties to transactions, countries that the customer is dealing with, and many more dimensions start providing rich information about the customer. This information can be utilized through statistics and can be converted into updating customer risk ratings.

There are various forms of applications. As a modeler, we can either convert it into a risk rating (every customer will have a probability to commit a financial crime in the next 12–24 months) to a dynamic segmentation—behavior-driven segments, that enable every customer to be classified in a particular segment and then monitored accordingly for the next quarter or six months. If we intend to do a risk rating, the choice of event would be possibly a binary outcome (whether a customer is a perpetrator of financial crimes or not). For segmentation, there are many classification algorithms that can be used including clustering. A clustering exercise will simply classify customers based on their transaction behavior and few static information like industry, banking relationships, etc., once the segments are defined, we can gain an understanding of the riskiness of such behaviors by identifying the concentration of perpetrators of financial crimes in those segments.

Definition of financial crimes could be money laundering, aiding tax evasion, and terrorist financing.

Once the scorecards are developed, they need to undergo regular model governance and back testing. These topics are again applicable to multiple modeling exercises from the perspective of a financial crime. They are explained in the appendix to avoid duplication.

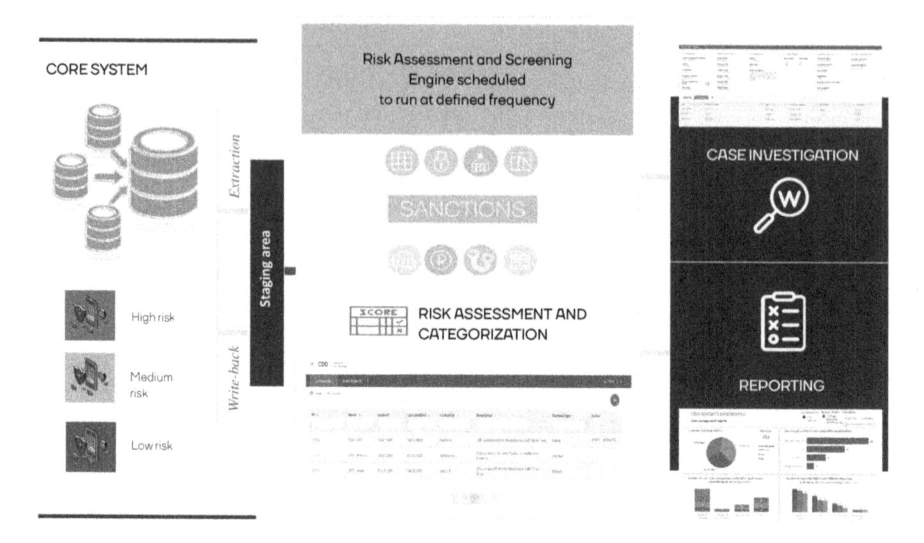

Fig. 6.5 Showing the process of automating customer risk assessment through application of machine learning model

6.5.4 Conversion of Machine Learning Outcomes into Smart AI Engines

We have mentioned multiple mechanisms for generating machine learning models for providing better visibility and intelligence into the customer risk assessment process.

As shown in Fig. 6.5, these need to be converted into solutions that provide this intelligence and decision on an automated basis. Once integrated into the core solution, they can optimize the process and improve decision-making while reducing the cost of running the process manually.

With the application of digitization and artificial intelligence, an end-to-end process can annually save a couple of million minutes of resources. With the increasing pressure from the business and regulators, compliance will have to adopt them. The only question is when and how prepared are they for making this transition.

Artificial Intelligence-Driven Effective Financial Transaction Monitoring

7

7.1 Introduction

Once the customer is onboarded, a logical progression is conducting financial transactions with others—be it other customers of the bank or outside entities. As we discussed earlier, the landscape and nature of financial transactions are increasing significantly.

While a financial institution acts as a financial intermediary for facilitating these transactions, some of the transactions entail transactions of laundered or illegal money. A very small portion of these transactions is also related to transfers done to sanction or watchlist entities. A financial institution is required by law to identify, block, and report such transactions to the regulator. This is one of the areas, where a lot of financial institutions have already been fined millions of dollars for failing to control this. This not only results in financial loss for the organization but, it also results in reputation risk for the financial institutions. Another implication is stress and time wasted for the senior management in physically being present and cooperating in the ongoing investigations.

With such a ramp up in the oversight and control by the regulators, financial institutions are increasingly becoming wary of facilitating such transactions. Having said that, identification and blocking/reporting of such transactions is a very labor-intensive and time-consuming exercise. As per the global payments report, the total number of non-cash transactions globally are around 785 billion in 2021. This is expected to grow to 1.8 trillion transactions over the next 5 years. Currently, based on a rough assessment of the global SAR filed, less than a quarter of a percent of such transactions are identified as suspicious and are investigated (Fig. 7.1).

Financial institutions have traditionally been monitoring such transactions through a set of automated rules that trigger an alert for such transactions. We

A. Gupta et al., *Artificial Intelligence Applications in Banking and Financial Services*, Future of Business and Finance, https://doi.org/10.1007/978-981-99-2571-1_7

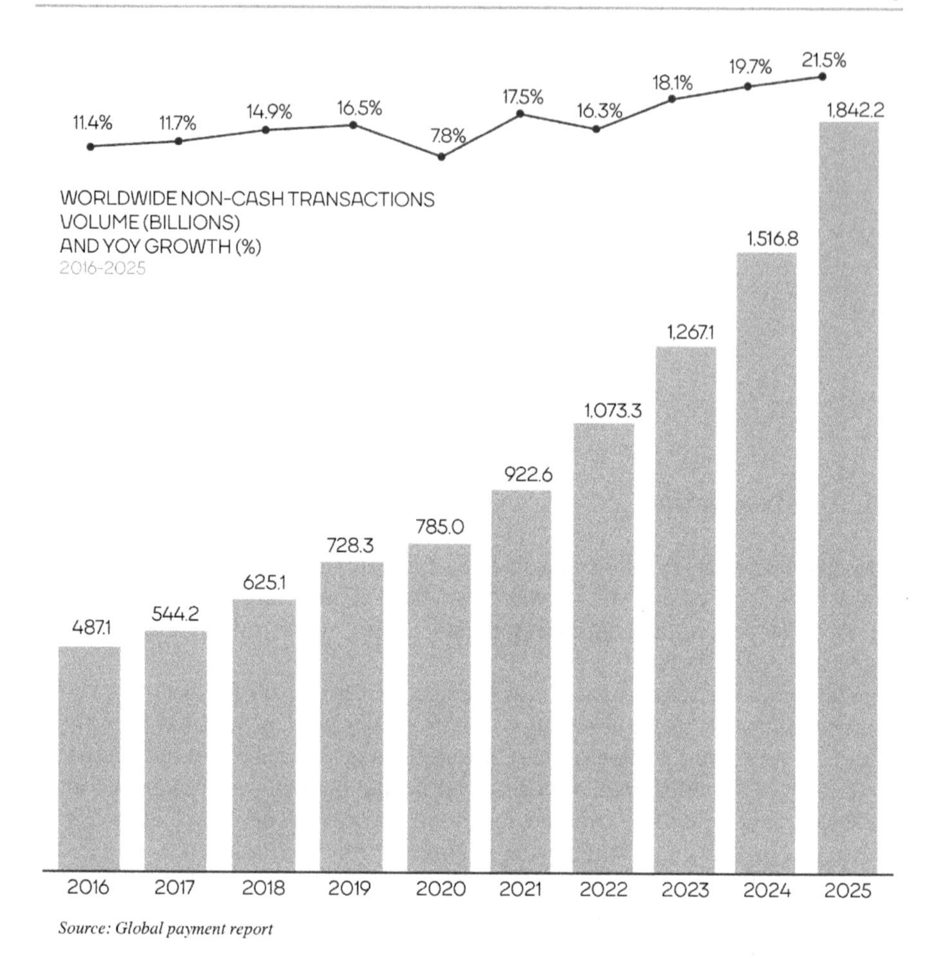

Fig. 7.1 Showing global non-cash transactions

have discussed examples of such rules in our previous chapter. While these scenarios (translated into one or set of rules) help the organizations in identifying suspicious transactions, they also result in many false-positive alerts.

To add to this phenomenon, a significant number of regulators are also recommending new areas of transaction monitoring during their audits. They cater to the new types of transactions, new customer segments, and emerging themes on financial crimes. FIs need to cope with this regularly expanding coverage while managing cost and time. This brings to us the concept of alert optimizations. Before we discuss various mechanisms of alert optimizations, let us first focus on various types of transaction monitoring:

(a) Real-time counterparty monitoring for countering financing to a sanctioned entity

(b) Scheduled review of financial transactions for money laundering.

(a) Real-time counterparty monitoring for countering financing to a sanctioned entity

Before a financial transaction is executed, a financial institution is supposed to check the counterparty and block the transactions, where the counterparty is a sanctioned entity.

Depending on the type of financial institution and the jurisdiction, there are also a few rule-based checks that are applied to the customer transactions on a real-time basis. For example, in the case of money exchange businesses or authorized money changers, there is a limit on the monetary transaction, till which the financial institution does not need to do customer KYC. But if the customer transaction value over the last x number of days exceeds a defined amount, then the financial institution cannot proceed with the transaction without KYC. These types of checks also are to be covered in the real-time transaction screening.

The most used real-time screening is the sanctions screening. It must be real-time monitoring, as financial institutions are promoting a higher degree of real-time payments. Another important aspect for European countries is also established ultimate beneficiary for the transactions. These once satisfied result in real-time settlement of the transactions.

Imagine a situation, where customer A has made a transfer of USD 500 to an individual B from another bank. Bank of customer A needs to ensure that this money transfer is not made to B, if B is an individual or entity, the bank of B or the country in which B's bank is located, is on the sanctioned list. The way the transaction will be executed is as explained in Fig. 7.2.

(b) Scheduled review of financial transactions for money laundering

Money laundering activities are perpetrated by individuals or groups of individuals following different methodologies. A financial institution that is intermediating and facilitating this transaction is obligated to monitor and report such transactions if they seem suspicious. These are monitored on an ongoing basis by financial institutions through scheduled transaction monitoring. During the discussion on scenarios in the previous chapter, we explained various forms and mechanisms of monitoring the transactions.

One of the important elements of a scenario is the parameters that are utilized to trigger an alert. The other important element is the threshold for those parameters. Let us first explain this through an example:

People doing money laundering typically rely on cash withdrawal and cash deposits. Let's say, one of the scenarios implemented looks at cash withdrawal. The definition of a scenario is a large cash withdrawal within a given timeframe. For such a scenario, the business rule will aggregate all the cash withdrawals for any given time frame. If the total amount of cash withdrawal is greater than the

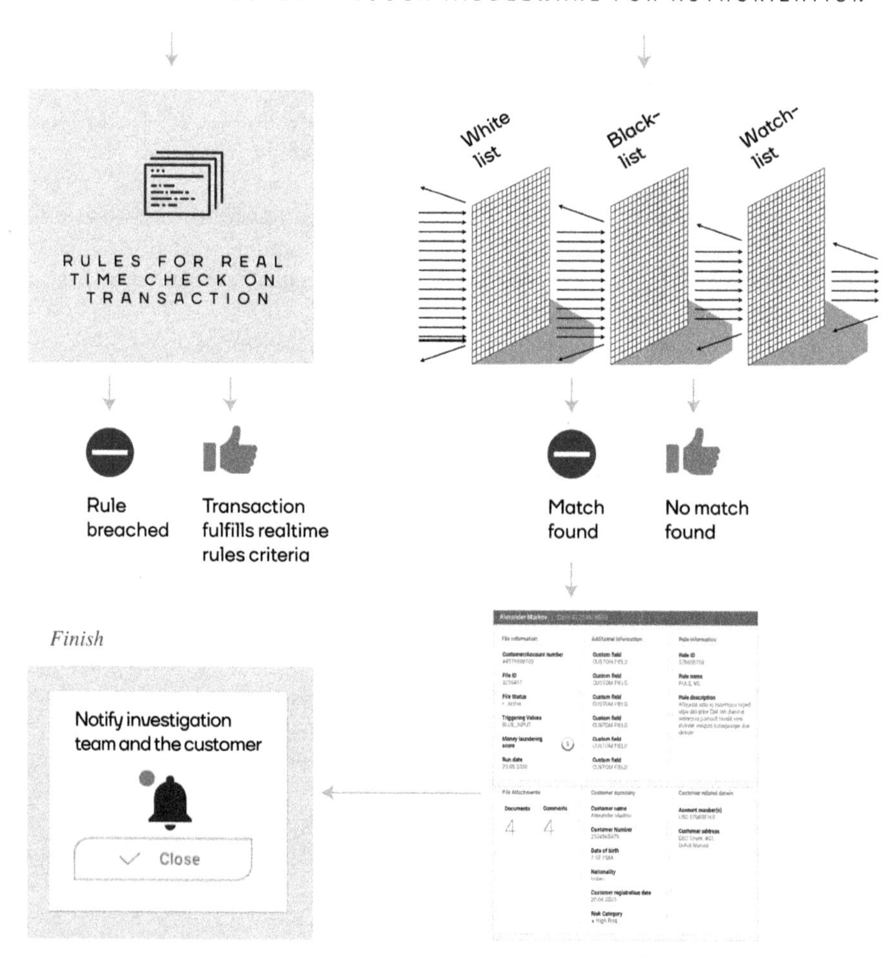

Fig. 7.2 Process for authorizing real-time transaction monitoring while initiating a transfer

threshold, it will generate an alert for the investigation team to assess, whether the transaction is bona fide or not.

The above scenario has two specific parameters—the number of days for which the cash transaction has to be aggregated and the total amount of cash transactions within those number of days. To generate an alert, the compliance team needs to define the threshold for these two parameters. Say it is defined as 1 day and USD 5000. It will mean that the scenario will assess the total cash withdrawal aggregate for all the customers within the bank. If the total withdrawal exceeds USD 5000 on any given day, the solution will trigger an alert for investigation.

While (a) and (b) above can trigger scheduled alerts depending on the definition of the scenario, we will recommend handling (a) with machine learning algorithms. (b) can also be handled through a machine learning algorithm. This will be discussed in detail in the next chapter. For now, we will focus on a specific aspect called threshold finetuning.

7.2 Threshold Finetuning

Most of the financial institutions are not financially savvy enough to develop and implement machine learning models for alert optimization. For them, threshold finetuning is an excellent way of handling such a challenge.

Threshold finetuning is an exercise that entails managing the thresholds for every scenario in such a way that it helps the FI in maximizing the coverage while minimizing the false positive. Regulators across the world are now recommending FIs to undertake regular threshold finetuning on an ongoing basis. Threshold finetuning has a few specific terms that will be discussed subsequently. They are:

Threshold—The value of any parameter in a scenario that is breached will be used as a trigger for the compliance team to investigate transactions done by a customer. In our above example, a cash withdrawal of USD 5000 for a single transaction is the definition of a threshold.

Alerts—An automated trigger of transaction/group of transaction that exceeds the threshold translate into an alert. Alerts are investigated and either closed with false positives or they translate into suspicious activity filing (SAR) with the regulator.

Case—Alerts are prima-facie investigated and closed as false positives in most of the cases. If the investigator believes that prima-facie, there exists suspicion on the mechanism of the transactions, the investigator creates a case. Alerts are not deeply investigated unless they are converted into a case. Investigator can also create a case and can assign multiple alerts from a single individual or a group of individuals, depending on the initial findings of the investigation.

Triggering transactions—A customer can do hundreds and thousands of transactions with a financial institution. However, there are a few transactions that eventually resulted in alerts being generated and investigated. The transactions that trigger the alerts are titled triggering transactions.

Productive alerts—Alerts have a false-positive rate of anywhere between 95 and 99.9%, depending on the quality of the scenario. The alerts which eventually translate into case investigation and a SAR filing will be called productive alerts. One challenge that we have faced with various financial institutions is that while doing SAR filing, they don't capture the alert ID or transaction IDs under which the customer has been given SAR filing. In that case, all alerts or alerts around anomalies are called productive alerts.

Productivity rate—The ratio of total productive alerts to the total alerts for any given scenario is called productivity rate. Threshold finetuning can have productivity rate maximization as one of the objectives. We will discuss that in the latter section of the chapter.

SAR coverage—This refers to the total SARs captured by the scenario. It is defined as the number of SARs captured by the scenario divided by the total SARs that are filed by the bank.

Customer coverage—A customer can do multiple suspicious transactions for doing money laundering. It is possible that a customer has moved a large amount of money through wire transfer and also withdrawn cash. This means that there could be multiple alerts for the same customer. In our analysis, we have started relying on an additional parameter called customer coverage. It basically refers to a customer-level metrics for productivity as against alert-level metrics.

Overlap ratio—Overlap refers to similar customers who are triggered by multiple scenarios. It helps in identifying the utility of scenarios for capturing risky customers, in the presence of other scenarios. We use this parameter for optimizing our patented algorithm for cross-scenario optimization. We will discuss this in more detail in this chapter.

Anomaly detection—The majority of the customers transact in such a way that their transaction patterns match either with the profile of the customer or their segment. These are called normal transactions. However, due to certain circumstances or the perpetrators of financial crimes, often do transactions that do not reflect their past behavior or of the segment that they represent. These are called anomalies. One of the objectives of investigators or data scientists is to look for those anomalies which can lead them to eventual money launderers.

7.2.1 Objectives of Threshold Finetuning

As discussed, threshold finetuning is not only a financial institution-triggered activity. The majority of the regulators across the world have prescribed regular threshold finetuning exercises. In this exercise, the compliance team conducts the exercise, along with the documentary evidence of threshold finetuning and their reasons. Some of the objectives of finetuning are:

- Mapping the evolved behavior of the customers—Customers and segments are not static. Their behavior evolves. Sometimes there are micropatterns due to channel usage, migration of channels, a fee levied, etc., while in other cases, there are macrofactors. For example, inflation reduces the purchasing power of money. So, over a period of time, thresholds need to be adjusted for the right reflection of the customer behavior
- Optimize the productivity rate of scenarios
- Try to have the analytical basis for setting thresholds rather than judgment-driven thresholds.

There exist multiple works of literature on threshold finetuning. Various consulting and technology firms have provided threshold finetuning as a topic to maximize the productivity rate. Productivity can be estimated at an alert and transaction level. For a scenario, the productivity will be defined as the total number of alerts that resulted in SAR or case creation. If one divides this by the total number of alerts generated within the scenario, it will be called productivity rate.

A similar concept can be applied for estimating productivity for transactions or customers. However, in general parlance, productivity is estimated at an alert level.

With an understanding of traditional approaches, objectives, and constraints, we would now understand the proposed approach (Fig. 7.3).

Segment definition

A precursor to the threshold finetuning is the definition of segments. Users who are exposed to an investigation already do this regularly. Which is called

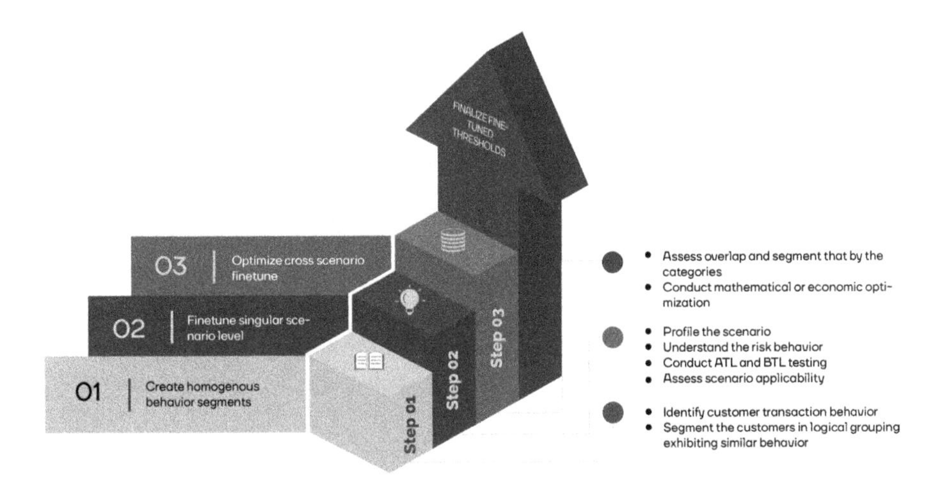

Fig. 7.3 Process for optimizing alerts through threshold finetuning

defining segments. All customers are not the same. Their backgrounds, reasons for banking, type of product and counterparties, and the turnover of the accounts vary significantly by customers. Keeping this in mind, the majority of the organizations rely on defining segments.

The traditional definition of segmentation is static segments. This methodology of segmentation considers company background, type of business, politically exposed persons (PEP), income/bank definition of customer segment based on the size of relationship, sometimes industry of the company, and other factors. This approach also works, given that they are generally not based on individual biases and segments provide the judgmental experience of customer risk from the perspective of a financial crime. However, it has its own challenges. Most important is a huge anomaly in terms of the size of segments. There will be segments that are very small making them impractical for any meaningful analysis. We have come across segments where number of customers within them are 15. Such a small segment does not aid in any meaningful analysis.

On the other hand, there will be significantly large segments comprising almost 40–45% of the portfolio. Another challenge with defining segments this way is that they do not encapsulate the actual behavior of the customer segment. A static segment is good when there is no information about customer behavior. When a bank can assess customer behavior, it makes sense to base segments based on dynamic customer behavior. Let us try explaining this heterogeneity within the segment through an example. Imagine an SME customer who is not engaged with its FI versus another SME customer who is completely relying on its FI for executing all transactions. These two customers, despite belonging to the same segment are very different in terms of their transaction behavior. A similar threshold will not reflect their true transaction behavior and hence will result in a significantly higher number of false positives.

Heterogeneity also poses a challenge in explaining the anomaly in their transaction behavior keeping the above in mind, a precursor to the threshold finetuning is segmentation. Our methodology relies on the unsupervised learning-driven classification of customers, based on their transaction behavior and their background. The process for the development of customer segments is explained as follows.

As shown in Fig. 7.4, the process of segmentation for any modeler is a combination of art and science. Factors to consider for the segmentation are as follows:

Ethical decision-making—factor selection for segmentation exercise can be biased—it could be based on ethnicity, geography, race, or similar factors. It is imperative that factors considered are reflecting the risky behavior of the customer, rather than the customer's demographic background. Based on behavior exhibited, if there exists certain explainability on demography it can be retained with due diligence from a business point of view. The reason is, that when segmentation was being defined, it was based on transaction behavior. Modeler did not even have an inkling that this could map to a particular profile of the customer, and hence the human bias is avoided.

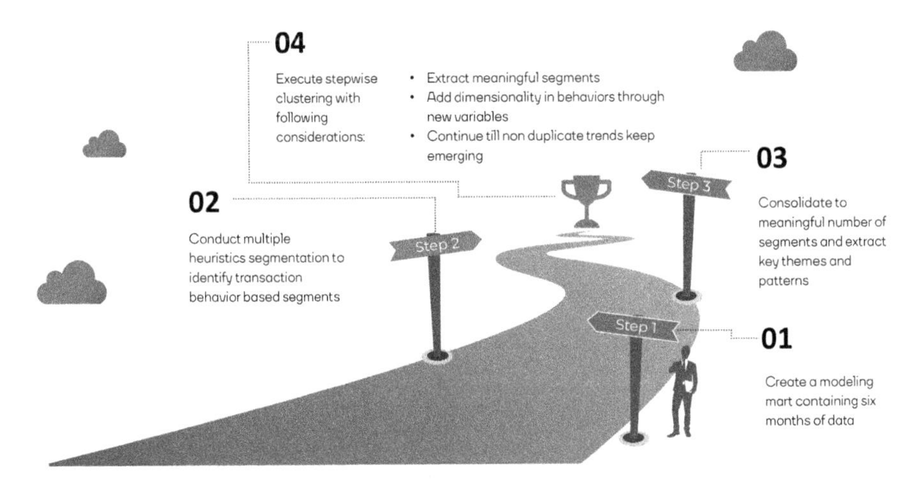

Fig. 7.4 Showing segmentation steps

The above is not an essential step for threshold finetuning. However, it helps to align these before conducting a threshold finetuning exercise (Fig. 7.5).

Our explanation of the threshold finetuning is explained as follows:

As explained in Fig. 7.6, we conduct threshold finetuning with twin objectives—maximize the alert and customer productivity rates while maintaining the overall SAR coverage for the customers.

The first stage entails analyzing the parameters of a given scenario. Before attempting a scenario threshold finetuning, the data scientist needs to understand the business intent, the parameters, and their threshold setting rationales for various

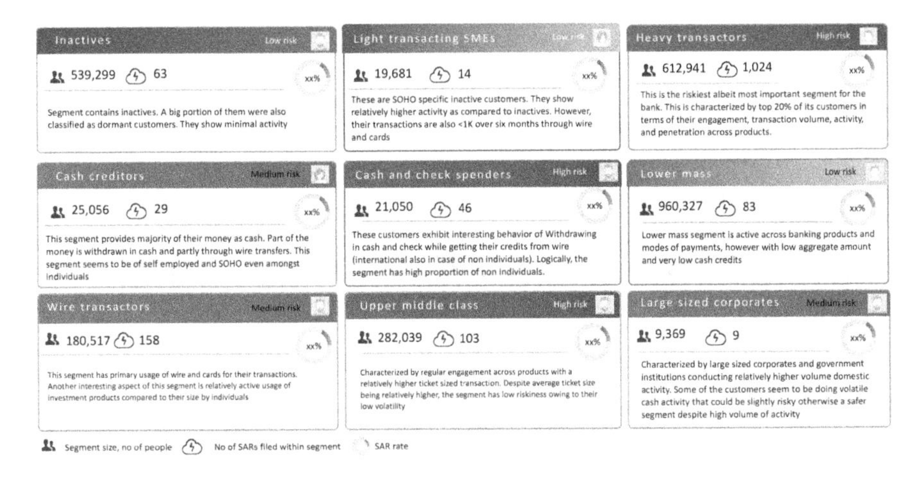

Fig. 7.5 Showing sanitized example of behavioral segmentation

ORIGINAL

Segment	Minimum ATM withdrawal	Minimum credit amount for last 5 days	ATM withdrawal to credit ratio	Total Alerts	Productive Alerts	Non-Productive Alerts	Productivity rate (%)
Corporates	1,000	1,000	90	92	1	90	1.09
HNI	1,000	1,000	90	76	-	76	0
SME	1,000	1,000	90	8,756	15	8,741	0.17
Individual - Mass	1,000	1,000	90	61,221	500	60,722	0.82
Individual - priority	1,000	1,000	90	14,919	47	14,873	0.32
SOHO	1,000	1,000	90	212	1	211	0.47
PEP	1,000	1,000	90	78	-	78	0.00
Commercial	1,000	1,000	90	19	-	19	0
	Total			85,373	564	84,810	0.007

ITERATION 1

Segment Id	Min ATM withdrawal today - Threshold 1	Min Total Credit period - Threshold 2	Min (ATM debit period/Total Credit period) Pct - Threshold 3	Total Alerts	Productive Alerts	Non-Productive Alerts	Productivity rate (%)
Corporates	10,000	50,000	90	37.0	1.0	36.0	2.70
HNI	10,000	50,000	90	30.0	-	30.0	0
SME	7,500	40,000	90	3,502.0	7.0	3,497.0	0.20
Individual - Mass	10,000	25,000	90	24,488.0	224.0	24,265.0	0.91
Individual - priority	11,500	50,000	90	5,968.0	21.0	5,947.0	0.35
SOHO	10,000	25,000	90	85.0	-	84.0	0.00
PEP	12,500	40,000	90	31.0	-	31.0	0.00
Commercial	10,000	50,000	90	8.0	-	8.0	0
	Total			34,149	253	33,898	0.007

ITERATION 2

Segment Id	Min ATM withdrawal today - Threshold 1	Min Total Credit period - Threshold 2	Min (ATM debit period/Total Credit period) Pct - Threshold 3	Total Alerts	Productive Alerts	Non-Productive Alerts	Productivity rate (%)
Corporates	10,000	100,000	75	1	-	1	-
HNI	10,000	90,000	30	46	3	46	7
SME	7,500	70,000	75	1,233	14	1,233	1
Individual - Mass	10,000	25,000	75	8,053	556	7,497	7
Individual - priority	11,500	100,000	75	9	-	9	-
SOHO	10,000	100,000	75	-	-	-	-
PEP	12,500	125,000	75	-	-	-	-
Commercial	10,000	150,000	75	-	-	-	-
				9,342	573	8,786	0.06

ITERATION 3

Segment Id	Min ATM withdrawal today - Threshold 1	Min Total Credit period - Threshold 2	Min (ATM debit period/Total Credit period) Pct - Threshold 3	Total Alerts	Productive Alerts	Non-Productive Alerts	Productivity rate (%)
Corporates	10,000	100,000	60				
HNI	10,000	90,000	30	7		7	
SME	7,500	70,000	60	17		17	
Individual - Mass	10,000	25,000	60	155	7	148	
Individual - priority	11,500	100,000	60	1		1	
SOHO	10,000	100,000	60				
PEP	12,500	125,000	60				
Commercial	10,000	150,000	60				
	Total			180	7	173	0.04

Fig. 7.6 Reflecting the productivity for ATL—BTL testing

segments. This is akin to developing a business understanding of the scenario that needs finetuning. Once these are clear, we move to the next stage.

The second stage entails replicating the scenario in our environment that has the flexibility of finetuning the threshold. It requires replicating the logic used to trigger alerts. Once the alerts are triggered, the team uses them to reconcile whether the team has understood the scenario correctly, as well as whether there exists the right data to compare the result. This step is also called baselining the finetuning the exercise. At this stage, all parties involved have a common understanding of what is the current productivity and coverage of the existing scenario. Any discrepancy is discussed and resolved, before moving to the next stage.

The third stage entails above the line (ATL) and below the line (BTL) testing. In simplistic language, it means modifying the threshold to understand the impact on alerts generated. It is commonsensical that if the threshold will be reduced, the number of alerts will increase and also possibly SAR coverage. Similarly, if the thresholds are increased, logically a number of alerts will fall and also the SAR coverage. The whole game of optimization hangs on maximizing productivity without compromising on SAR coverage significantly. In many cases, teams need to generate multiple iterations as the combinations of threshold parameters can be many. Let us try to explain this through an example.

Figure 7.6 takes an example of a high-velocity ATM withdrawal. The scenario captures situation where a customer receives credit in the account through any mode and use ATM withdrawal over last five days to take the money out. The client while implementing this started with default value of USD 1000 for all the parameters as threshold. While it is important to understand this sensitivity, one will realize that there is not an easy answer to define an optimal point. Mathematically, it would logically entail that with an increasing threshold, one would miss a few opportunities to catch money laundering events. By reducing the threshold, it would result in a higher catch rate of money laundering events, albeit with higher inefficiency of a lot of false positives.

The objective would be to understand correlated scenarios and then work on the scenarios in tandem to ensure that while true positives are not missed, productivity is enhanced.

As is shown in an example, the iterations are generated for a given segment. It becomes a joint exercise between compliance department and the data analysis team to arrive at a finetuned value. Once finalized, both parties can jointly document the original metrics and improvised metrics that forms core of the recommended finetuning.

7.2.2 Cross-Scenario Optimization

Readers would have realized by now, that threshold finetuning can already deliver a value in terms of making scenarios better applicable to each of the segments and thereby improve the productivity. We have further improvised this and have filed a provisional patent in this regard.

Before we explain the cross-scenario optimization, let us remind the readers about a term, we introduced at the beginning of the chapter. The term was titled overlap ratio. This essentially means that a customer doesn't perpetrate financial crime with a single type of transaction. There exist multiple triggers, and hence, there also exists overlap of alerts and customers across scenarios. This is precisely what we have tried to optimize through cross-scenario optimization.

For more details, a reader can also read our research paper (Abhishek Gupta et al.) that leverages linear or nonlinear programming to mathematically reduce the slack or overlap.

Tamás Badics et al. introduced another concept for optimization for overlapping scenario-driven alerts. The team leveraged a microeconomic model by transforming the threshold optimization into a decision problem that can be optimized through the economic evaluation of properties of the optimal point. Their recommendation is a weak mode estimation of the distribution for the algorithm. It can be extended to alerts and cases as well.

One last consideration with the threshold finetuning that needs some improvement from a data perspective is the identification and tagging of the behavior that led to the suspicion and eventual filing of a SAR. In our experience, financial institutions have not been able to clean this part of the data.

One of the potential problems with this type of data is that it is possible that the methodology that the money launderer used to launder money is to use alternative channels like multiple digital wallets and cryptocurrency to launder money, while his cash deposits were completely normal. In the absence of proper tagging of triggering transactions/behavior, even these are classified as productive transactions as they are tagged to the same customer. During threshold finetuning of the cash behavior, these will still confuse the data scientist in identifying true behavior and then eventually optimize.

As shown in Fig. 7.7, we have created synthetic data for transactions. Using that data, we have triggered alerts for a sample of five scenarios. As shown in the tables, the scenarios have a fair degree of overlap. In mathematical terms, we can call it slack (redundancy for a linear programming problem). This slack can be reduced by applying an optimization algorithm. Our results show that alerts can be further reduced by reducing this slack.

There are many mechanisms for example linear and nonlinear programming being two specific optimization techniques that can deliver this benefit.

Number of alerts pre and post optimization - example matrix

Unoptimized	CASH DEPOSIT	LARGE SPEND CARDS	INWARDS WIRE	OUTWARDS WIRE	AGGREGATE TRANSACTION	WEEKLY AGGREGATE	*TOTAL*
CASH DEPOSIT	3,132	297	121	347	2,223	891	*7,011*
LARGE SPEND CARDS	297	1,277	71	131	494	396	*2,666*
INWARDS WIRE	121	71	298	275	280	247	*1,292*
OUTWARDS WIRE	347	131	275	976	916	601	*3,246*
AGGREGATE TRANSACTION	2,223	494	280	916	8,736	3,017	*15,666*
WEEKLY AGGREGATE	891	396	247	601	3,017	6,855	*12,007*
TOTAL	*7,011*	*2,666*	*1,292*	*3,246*	*15,666*	*12,007*	*41,888*

Optimized							
CASH DEPOSIT	3,049	295	117	331	2,201	880	*6,873*
LARGE SPEND CARDS	293	1,225	68	128	472	387	*2,573*
INWARDS WIRE	117	70	286	271	273	241	*1,258*
OUTWARDS WIRE	331	130	266	937	912	583	*3,159*
AGGREGATE TRANSACTION	2,193	482	268	872	8,533	2,905	*15,253*
WEEKLY AGGREGATE	850	392	242	600	2,897	6,566	*11,547*
TOTAL	*6,833*	*2,594*	*1,247*	*3,139*	*15,288*	*11,562*	*40,663*

Fig. 7.7 Showing the reduced overlap among scenarios through cross-scenario optimization

8 Machine Learning-Driven Alert Optimization

8.1 Introduction

In the previous chapter, we have already given an overview of transaction monitoring. We have divided these into real-time transaction monitoring and scheduled financial transaction monitoring for money laundering purposes.

The former falls in the realm of sanctions screening. The machine learning-driven optimizations are already discussed in Chap. 6, where the name-matching algorithms and their applications in terms of real-time sanctions screening are discussed.

This chapter will focus more on machine learning-driven alert optimizations. We have advised several clients on machine learning-driven alert optimizations. The engagements have a few learnings in this regard. Most important is that machine learning-driven insights and its application will go through a journey. Be realistic about the value that is delivered. Understand, what can improve this insight and work on a plan to continue improving. This chapter will focus on various approaches within machine learning algorithms that could be used.

As shown in Fig. 8.1, there are different stages of sophistications, one can achieve through the modeling process. Reason for the first and second stages is also the data availability. AML teams have been collating transaction level and customer demography data to some extent for their alert generation process. However, information which captures 360-degree transactions, customer data beyond alert generation process, counterparty data of the transactions are not easily available.

Keeping the readiness in mind, the best way for an organization to start the AI journey is to develop generation 1 models with what they have. And eventually scale up.

A. Gupta et al., *Artificial Intelligence Applications in Banking and Financial Services*, Future of Business and Finance, https://doi.org/10.1007/978-981-99-2571-1_8

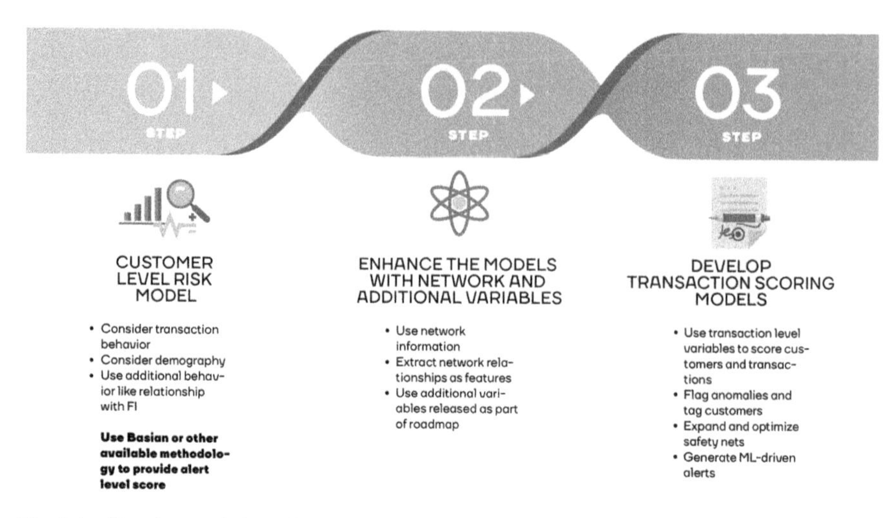

Fig. 8.1 Showing evolution of modeling approaches for alert optimization

8.2 Various Approaches for Assessing Customer Riskiness Through Machine Learning

While we have recommended a mechanism for the customer risk assessment, there also exists another mechanism. A summary of other mechanisms is provided below. A modeler can decide on any possible methodology or modeling framework to predict outcome. An important element would be to follow robust modeling framework to generate insights-driven predictive outputs. There could be gaps due to the selection of modeling process also, but that would need a discussion between business and data scientists to ensure that the approach is commonsensical and would give necessary insights that the business is looking for.

C. Suresh, K. T. Reddy, N. Sweta have suggested a mechanism to identify the path of the money laundering using a hash-based association so that one can automatically identify agent and integrator using graph theoretic approach.

Chih-Hua Tai, Tai-Jung Kan have proposed two-step approach where the first step entails identifying suspicious money laundering accounts. The next step works on retrieving more suspicious accounts to improve the precision. Their work done on Bank Sinopac data achieved recall rate of 26.3% in the first step and precision rate improves to 87.04% in the second step.

A. Shokry, Mohammed Abo Rizka, N. Labib have proposed an unsupervised machine learning technique to identify money laundering transactions and groups for counterterrorism. They worked on comparing two algorithms and then selecting the best algorithm.

Nhien An Le Khac, Sammer Markos, M-Tahar Kechadi have applied data mining techniques on investment data. Their contention is that accounts, transactions, and institutions or individuals doing them are interrelated. They combined them to create groups. Then the transaction behaviors were mapped to create profiles. This profile was mapped to classify customers into pre-defined categories of risk. Essentially, it was a classification approach.

C. Suresh, K. T. Reddy have used heuristic approach in the form of decision rules for conducting the link analysis. To bring efficiency in large volume of transactions, they also leveraged multitable joins through bit-mapped join indices that helped in reducing processing time.

U. Ketenci et al. developed a new framework which is called time–frequency analysis. The premise of this approach is that suspicious behavior and otherwise can be mapped on time form. It can be converted into a two-dimensional representation of transactions with time as one dimension and frequency as another. Through this profiling, the suspicious and non-suspicious behaviors can be mapped and classified.

As discussed, there can be multiple forms of modeling approach. After the due selection and syndication, data scientist can proceed with a selected approach. Let us proceed with our recommended approach.

We will start with the simplistic form of machine learning-driven alert optimization.

8.3 Customer-Level Monitoring and Scoring

As explained earlier, money laundering is done by a customer or a group of customers with a common objective. In these cases, they use multiple mechanisms and modes of transaction behavior, channels, geographies, and counterparties. A machine learning algorithm provides an improvement over alert-level framework, as alerts are unidimensional; i.e., alerts can encapsulate a particular transaction behavior at a point in time. Machine learning algorithms provide a lift on predictive power by combining multiple dimensions in a time frame. Not only that, but machine learning algorithms can also look back into the behavior of the customer in the last few months to then define anomalies in transaction behavior and hence the riskiness.

The process for alert scoring is provided as follows:

In this case, a big portion of work goes into the development of a customer-level risk score for every week. The way to achieve this is through developing a model mart that captures week-over-week data for the last 3–6 months. A data scientist may want to develop even a daily mart. It depends on the dynamism of the financial transactions and investigations. We had to rely on last 6 months, since the triggering transactions and subsequently investigation to SAR filing cycle took almost 3–6 months. We have already highlighted this issue in our data quality section in Chap. 4. It has also been cited as an issue in one of our papers titled "Data quality issues leading to suboptimal machine learning for money laundering models" in the Journal of Money Laundering. Another mechanism that is adopted

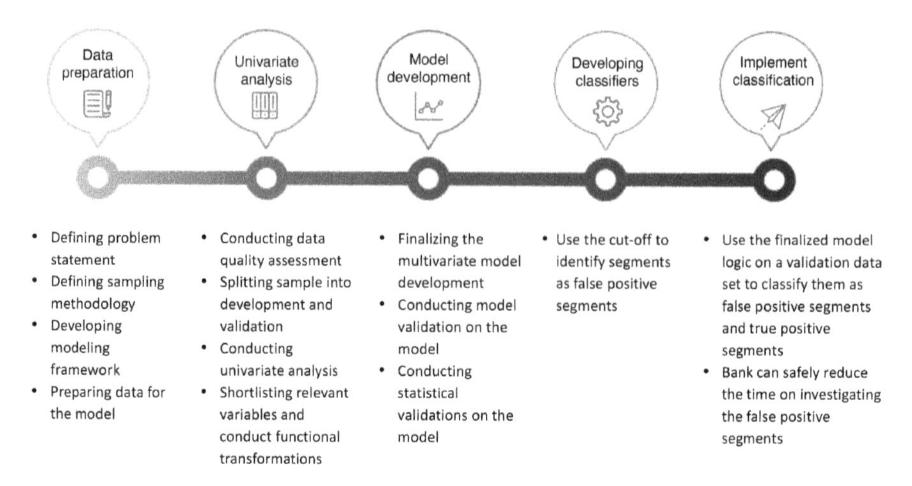

Fig. 8.2 Providing an overview of the modeling lifecycle

by the financial institutions is to use only the alerts to train the model for alert scoring. Each method has its pros and cons, however, both are valid methodologies.

Modeling lifecycle

As shown in Fig. 8.2, a modeling lifecycle is divided into five broad stages. Four of them are directly linked to the model development lifecycle. The last one is about converting the process and logic into steps for the technology team to implement. With the advent of new technology, these two process elements, i.e., development of model and deployment of the model, could be merged together.

There are various products available in the market that has the functionality of model development and then converting the model development lifecycle into a package. This package can be directly exposed in the form of APIs. Thus, a big chunk of efforts of model deployment and documentation could be saved through adoption of those packages.

Coming back to model development lifecycle. Let us go over four stages of the model in the context of alert scoring.

Problem Definition and Data Preparation

Some of the important elements for the development of such a model are:

Definition of event—for the development of such a model, we consider all customers for whom the SAR is filed as the definition of event. In the cases where the number of SAR filings is not sufficient, then the customers for whom the case investigation is also done are covered under event. It basically refers to confirmed money launderers and highly suspected money launderers as our target group, whom we will use to predict.

Please bear in mind, that SAR filing can happen because of the following reasons:

1. Enterprise AML solution-driven triggered SAR filing
2. Central bank initiated SAR filing where Central bank asks the FIs to provide information on SAR customers
3. Adverse media-driven SAR filing.

Among the above, only (1) is considered as relevant for definition of event. Others may or may not have sufficient transaction history—imagine a customer who has an account with a particular FI but is dormant on this FI. Just because he has been investigated by the central bank for suspicious activity, doesn't give any reason to the current FI where he/she is dormant. If a modeler will end up using such cases, it will result in poorly defined modeling framework and weaker model that will have high classification error.

Predictor Variables

A modeler needs to conduct multiple workshops with the business teams to identify all possible factors and typologies that can give any indication of the money laundering event. Each of these needs to be translated into a data request and then eventually an exercise to map this data should be done.

In case, FI does not keep a sufficient history or if the data is not available, then those wish list variables can be excluded; however, the attempt from the modeler should be to maximize the number of variables. Some of the dimensions for the variable creation are discussed in Chap. 4. Here we will discuss some of the common risk typologies. These are relevant typologies; we have come across different geographies:

- Fraud-related events, that can be induced by any type of fraud
- Virtual currency-related transactions
- Tax crime conducted by individuals and organizations
- Corruption and bribery
- Drug and human trafficking
- High-velocity transactions coming from high-risk countries. Analyze the portfolio of the organization to identify high-risk countries. Sometimes they happen to be non-prime suspect countries. FATF prescribed list can be a good starting point
- High-risk industries like casinos and gambling, real estate, jewellery, construction-related services, and specific retailing concepts
- Increasing usage of payment services
- Transfer of funds for golden visas
- Organizations started by immigrants from specific countries, whose net worth doesn't match with the company turnover and their stay in the country is very short
- High amount of cash transactions
- Transaction behavior not matching with KYC
- Payments from crypto-accounts, not matching the customer profile.

There can be many more such typologies depending on the organization. An important step is to map and understand these. Also keep an eye on emerging ones as in many cases, the lifespan of a few typologies is short, and they keep emerging.

Modeling Horizon

A decent size organization typically has anywhere between 200–2000 annual SAR filings for a country. Depending on the availability of the data on SAR customers, a modeler can choose the modeling horizon. We typically recommend the last 2–3 years as modeling horizon.

It is done with the following objectives:

- Relatively longer time frame exhaustively encapsulates various ML typologies. Data analysis hence provides sufficient coverage
- It ensures relatively large number of SAR customers to develop a statistically robust model
- Selecting further longer modeling horizon translate into slightly outdated data which can weaken the model. It implies that if an FI can have longer horizon of transaction and SAR data, then the data analyst should look for applicability and transaction trend. Choice of extended time frame can be done based on these points.

Observation Window and Performance Window

For any modeling exercise, one of the important elements to manage is the definition of the observation and performance window. In an idealistic world, the date on which SAR is filed can be considered as the beginning of the performance window. Depending on the history of anomalic behavior, the modeler can define the observation window. The observation window can vary from anywhere between 1 and 6 months.

In the real world, there is a big issue of the timing of SAR filing. There are cases, where the SAR filing is done after almost 6–9 months after the triggering events. Another challenge in the real world is that the FIs provide the list of SAR-filed customers. They do not have an approximate date of triggering transactions. If we literally take the SAR filing date as the performance window, then there are customers, who have no transaction history in the last six months and that can result in a poor-quality model. In such a case, we conduct data analysis for the last 12 months. We identify potential anomalic behavior, which could be considered as the beginning of the performance window. Accordingly, the observation window for the transactions could be determined.

Once the problem definition is completed, the next stage for the model development is the data preparation.

From the model mart perspective, a detailed explanation of data quality steps and model mart preparation is provided in Chap. 4.

One important lesson in this process is the time spent by the modeler in understanding various kinds of transaction behaviors. The more one gets into the details, more one gets to create and test transactions behavior. This brainstorming or hypothesis generation defines the quality of model alongside other important aspects of the model development process.

Univariate Analysis

Once the model mart is prepared, the next stage for the model development lifecycle is conducting univariates. However, an important step before any analysis is to split the sample. This is important to ensure that the model developed is not overfitted. Overfitting in simple words means that model has captured the inherent trend provided in the data very well. This is because the modeler ended up specifying every specific behavior as an input into the model. The problem with being very specific about even exceptions is that model will start predicting even the exceptions. The outcome will be a highly predictive model on the sample on which the model is developed. When the model is deployed on a portfolio or in real life, those over specifications do not exist. Hence, the model performance drops significantly. To ensure that the model is indeed working, the modeler should always keep aside a randomly selected portion of data safe without bringing any bias in the splitting process.

Once the data is split, it will be divided into a development sample or training sample/validation sample, and in a few cases, out-of-time validation sample.

The development sample is the dataset that will be carried forward for model development. Modeler needs to start conducting univariate analysis. While an auto ML models would be able to do this exercise in an automated manner, a human intervention can also identify some of the multivariate trends which machine may or may not be able to capture without unsupervised learning.

This step entails analyzing the trends, patterns derived variables and correlating them to the outcome. In our case, it is SAR or case creation. This analysis helps modeler in providing insights around the nature of relationship different predictors have with the outcome. It also provides insights on the linearity of the relationship of predictors to the outcome.

For categorical variables like industry, the modeler will be able to group these discrete industries into homogeneous groups based on their riskiness. This process is called binning (Fig. 8.3).

Shortlisting the variables

Depending on the breadth and type of transaction behaviors, a modeler is left with choice of plentiful. There can be quite a few variables having very similar types of strength in terms of predictive power.

Another challenge within similar predictive variable is the dimension, they represent. Let us explain it with an example. There are two variablcs that have similar predictive power—Total number of credit transactions in the last 4 weeks and total number of credit transactions in the last quarter. It is possible that both the variable

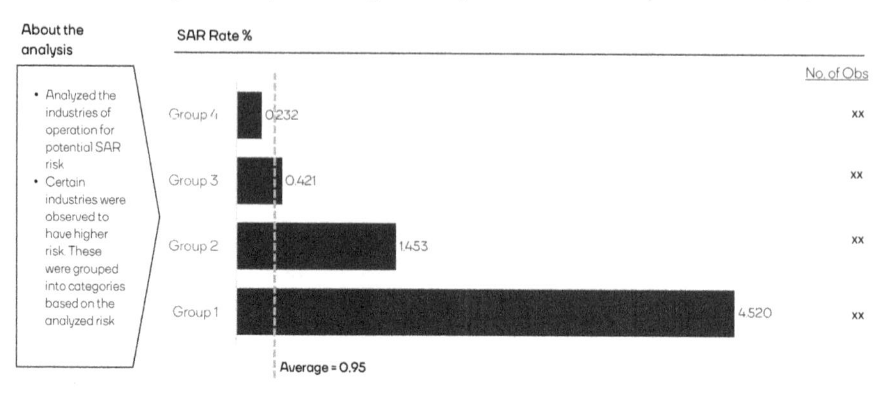

Fig. 8.3 Shows an example of a binning for variable creation

has relatively higher standalone predictive power as compared to another variable titled volatility of total payments in the last 4 weeks.

It is imperative that the modeler should know that the first two variables represent same dimensionality—credit transaction amount. It will create a problem of multicollinearity in the model. Also, model would be blind to a dimension that could be interesting, even if relatively less predictive, i.e., volatility of the transactions.

To handle this challenge, one need to either look at such correlations or dimensionalities manually. An alternate would be to run a factor analysis / principal component analysis that helps in organizing the factors based on their dimensionalities.

That helps in reducing the dimensionality and shortlisting the variables.

Functional Transformation of the Variables

In the univariate analysis, the relationship of predictor to the event is established. Categorical variables have already gone through binning and that can be converted into either transformed variables or a dummy variable. These are two broad ways to transform categorical variables.

For continuous variables, there are a few options available—(a) plot the independent variable on the x-axis and the dependent variable frequency on the y-axis. Use best-fit equation to be used for transformation (b) another option is to discretize the continuous variable. It could be either percentile range, judgment-based ranges, etc., and then use binnings to transform the variable and (c) use the raw value as is after ensuring it is treated for outliers.

The outcome of the functional transformation would be variables transformed value which is ready to be inputted into the model.

Multivariate Model Development

The final stage of the model development is multivariate model development. This stage requires shortlisted variables to be inputted into the statistical technique to provide an outcome—a predicted value that determines the riskiness of the transaction or the customer; whatever the case may be.

The choice of modeling technique and their pros and cons are discussed in detail in the appendix.

Depending on the choice of the modeling technique, the outcome varies. Depending on the technique, it can be a segment, a probability, or a score. The data scientist needs to convert this outcome into business inferable output.

Score Calibration

The outcome of the model is the probability for any customer to commit a financial crime, given his/her history of financial transactions and other markers. However, the probability might be difficult for business users to infer. Keeping that in mind, we convert the probability into a score.

The development of a scorecard entails converting the probability or the odds to a score. A score can be made by mapping odds to a score with a concept of distance to double the odds. For example, if the distance to double the odds is 30, then at every score sliding by 30, the odds to commit a financial crime will double. This can be designed in such a way that the investigation staff has an approximate idea of what score bands are riskier and which ones are relatively low risk.

Conversion of a Customer Score to an Alert Score

One of the expectations that we have seen from the business is that they want to continue using the existing process of alert investigation. Generally, the expectation is to provide minimal disruption to their process. We handle it through percolating down the customer-level scoring to the alert scoring (Fig. 8.4).

There are two ways of using the score—(1) A typical case investigation happens at a customer level. Hence, the alerts can be aggregated as a customer ID level. The score can be used to classify these groups as high risk, medium risk, and low risk. Eventually, the queues for the high risk and medium risk can be prioritized for investigation.

(2) The other variant of this is to convert the customer-level risk score applicable for the week or day to the alert-level score. Once the score is available for every customer, it is percolated down to the alert level.

Stage 2—application of social network analysis in improving the predictive power of the model

It has always been said that a human is a social animal. This is applicable to all including the perpetrators of financial crimes. A criminal rarely acts alone. It implies that if we can capture the trail of the path traversed by the criminal, we may be able to identify the criminal also. This is achieved through an evolved technique called social network analysis:

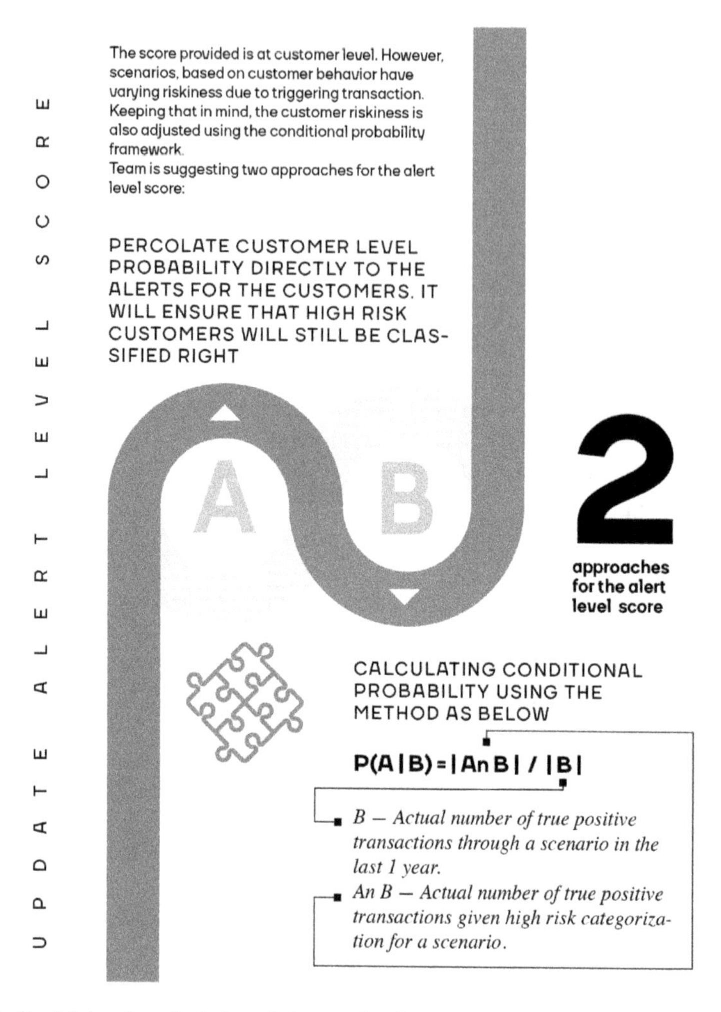

Fig. 8.4 Explaining the calculation of alert scoring from customer score

Before we go deeper into the technique, let us clarify that we do not recommend any organization to jump into social network analysis upfront. In our experience, the first-generation models capturing 360-degree view of the customer to generate a customer-level risk rating or alert-level risk rating, itself has its own challenges.

The quality of data, the ability of the compliance department to comprehend the results, the ability to consume the outputs, and then measure impact. All these activities are a change management program for any organization. If we open another front while we cope with bigger challenges, it runs the risk of too high an expectation with too many challenges to overcome. It could result in the program failing. In fact, in our experience, even first-generation models take almost a year to stabilize and improve as the quality of data keeps improving through its lifecycle.

Let us look at a couple of examples where network analysis could have raised alarm much faster. There is a group of individuals from a local salvage company who were cashing the checks all below CTR of USD 10,000. Eventually, it turned out to be a case of stolen chips from a computer manufacturer. Those chips were sold to the people in the salvage company. The employees of that salvage company were converting the proceeds from this stolen item through checks from the salvage company. Eventually, the network of owners and employees cashing checks at various intervals is caught after the aggregate amount of transactions exceeded significantly.

In the example above, while the triggers happened eventually as the total amount of proceeds translated into millions of dollars. A network analysis that is created in a short span and shows a significant spike, anomalic to their group behavior could easy generate alert which could be detected a lot earlier.

Having explained the context and timing, let us now go deeper into social network analysis.

Be it structuring, or simplistic velocity-driven transactions, the customers can create different types of networks among themselves for financial transactions.

We were recently conducting an analysis on networks and realized that a big chunk of SAR customers has a tight network among high-risk customers. Converting them into visualizations for the investigator would also be discussed in Chap. 9. For this chapter, it is about leveraging machine learning to identify, capture, and utilize from a network perspective.

Node—nodes are the objects or entities who are the part of the network. We are looking for high-risk entities.

Centrality—this has many measures but, in our context, the relevant centrality measures are the radial ones. More specifically the number of steps to connect high-risk nodes and the shortest-distance nodes.

Link—the nature and degree of association, as well as adjacency, can be assessed that can help analysts to unearth relationships of high-risk parties.

These derived values are captured and then inputted into the model. Our R&D on modeling techniques reveals that network-related variables provide a completely new dimension to the predictability of financial crime. They can improve the modeling power.

This is the reason; we recommend enhancing the power of network analysis for the prediction of financial crimes.

Stage 3—development of transaction-level model

This is the ultimate stage of the models. In this case, we have created data model similar to that of a fraud framework. Every transaction has a profile which contains historical trail of transactions. We are then using those trails of transaction to predict the risk of financial crime at a transaction level.

There are two possibilities for these models:

(a) Multivariate model containing various dimensions of transaction behavior into a single model. In this case, every transaction can literally be scored to a risk level
(b) Different multivariate models that track different type of transaction behavior like velocity, cash, multiple locations and so forth.

Both the modeling approaches have their own pros and cons. However, both provide improvement over existing alert-driven methodologies. We propose these transaction-level models to eventually replace alert generation at a scenario level. This will significantly reduce the false positive. It will also automatically capture the changing and evolving behaviors, thereby reducing the need to keep adding more and more scenarios in the alert-driven framework.

Another thing to recommend is to additionally provide safety net rules that should be appended to the models. There could be certain behaviors which drive the financial crime, but they are not statistically significant to appear in the model. To still cater to those, we would recommend, keeping an eye on those specific trends or behavior and then recommend additional rules to eliminate oversight from such behaviors.

These if followed in a reasonable manner, along with improving quality of data and implementation of these models in close coordination of business. These can deliver significant value in terms of high degree of false positive reduction, while improving the coverage of undetected money laundering events.

Another thing that can also be done to define a few safety net rules is to run unsupervised learning models on transaction behaviors. All anomalies are compiled and then converted into safety net rules. In our experience, they have significantly enhanced the power of compliance department in monitoring, detecting, and reporting such incidents proactively.

9 Applying Artificial Intelligence on Investigation

9.1 Introduction

One of the most important areas for AML, transaction monitoring, and sanctions screening is the investigation. Investigation provides the next big lever as scope of improvement in terms of efficiency.

For any financial crimes department to focus on efficiency, it comes from the reduction of false positives. The next area of optimization is the investigation. If you recollect the process of sanctions screening or the financial transaction monitoring, once an alert is generated, it needs to be investigated to ensure that either this is indeed a true-positive case or this is a false alarm. Irrespective, the case requires investigation which is a time-consuming and laborious process (Fig. 9.1).

A lot of FIs have leveraged bots that automate a few of the areas. However, we believe that the value can be enhanced for creating impact. The value lies both in terms of generating automated insights and reducing reliance on manual extraction, and compilation of supporting facts that help the investigator. Another aspect to keep in mind is the high turnover in the department.

Typically, the department is always looking for well-trained experienced professionals. With the employee turnover, the knowledge of the department gets diluted. Less trained professionals take more time and do more iteration during investigation, thereby increasing efforts and productivity of investigation staff.

With these aspects, we would focus on the information needs that aid an investigator both in terms of insights and automation of tasks. Some of the challenges we have seen in this department are:

- Scattered information thereby taking time in consolidation
- Manual generation of insights that slow down the investigation process
- Relying on manual media searches, manual collection of KYC information, developing profile of customer
- Differential knowledge level of professionals, thereby.

A. Gupta et al., *Artificial Intelligence Applications in Banking and Financial Services*, Future of Business and Finance, https://doi.org/10.1007/978-981-99-2571-1_9

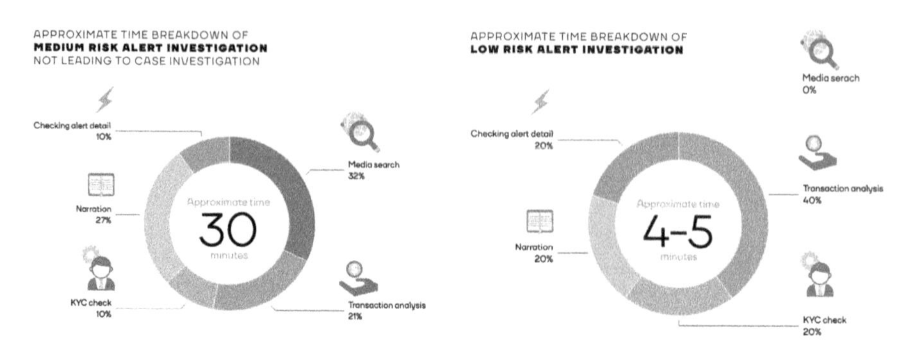

Fig. 9.1 Showing typical time spent by investigator on activities

In terms of improvement areas, we would be driven by the needs and efforts of an investigator. Hence, the opportunities are divided into two parts:

- Information needs of an investigator
- Automation opportunities.

9.2 Information Needs for an Investigator

An investigator is expected to investigate different aspects of customer transactions to ascertain whether the customer has tried doing anything suspicious. If prima-facie there exists a case for detailed investigation, then the investigator will create a case. For information needs, we have mentioned a few important dimensions of information:

1. Triggering transaction
2. History of alerts
3. Transaction profile
4. Debit profile
5. Credit profile
6. KYC information
7. Adverse media
8. Customer network
9. Narration for case investigation.

1. Triggering transaction

This functionality provides information on the transactions that resulted in an alert being triggered. In the case of the alert generation process or alert scoring process, the alerts will always have underlying transactions that result in their trigger. For an investigator, the first thing to check is those transactions. Analyzing them provides prima-facie clarity to the investigator, whether the alert is a false alarm or there exist sufficient reasons for the investigator to investigate.

Another important aspect of the investigation is the account-level and customer-level view. It is imperative for the investigator that the customer-level view is automatically consolidated. Most of the case management solutions provide a customer-level roll up. For example, if a customer has four accounts and a customer has attempted some suspicious activities in two of them, then there will be alerts generated on transactions done through multiple accounts. Depending on the scenario, an alert can be triggered by a single account or the sum of transactions done by the customer at a customer level.

Once an investigator eyeballs the consolidated transactions that have resulted in an alert being triggered, this coupled with a quick check on KYC profile and history of alerts can provide a quick judgment to close the alert as a false positive or further investigate it.

2. History of alerts

History of generated alerts is another important aspect for the investigator. History provides relevant inputs to the investigator in terms of the nature of alerts and activities which are investigated already and justified. Situations where the customer continues to transact on activities that were suspicious but not solid enough to file a suspicious activity report (SAR) in the past. All these inputs on historical alerts of the last six months can provide valuable insights into the investigator.

3. Transaction profile

The transaction profile provides a bird's eye view of the transaction behavior of the customer. These are analytically derived metrics that provide insights into the evolving transaction profile. There are metrics like turnover month over month, and any abnormal variation observed in transactions. They can also be broken down into transaction modes like cash, check, wire, and credit cards.

These can also be benchmarked to KYC information and industry peers. It helps in assessing whether the transactions done by the customer are in line with his profile, peer profile, and industry profile or they are indeed an anomaly, which requires a closer look.

This indicator can have a few indices like abnormality index for self or peer industry averages that provide a quick overview to the investigator and eases their task in making decisions.

4. Debit profile

This is an extension to the overall transaction profile. As the name suggests, it provides an overview to the debit transactions that a customer is doing. They are broken down into modes and products like cash, check, wire, credit cards, and digital wallets.

This enables an investigator to view the evolution of the debit transactions, and any major abnormality observed. There can also be a few analytically derived variables capturing the trend and the volatility.

5. Credit profile

A credit profile, like a debit profile, provides similar information. The only difference is that it provides an overview of the credit profile of the customer. It will also have its necessary drill down to ensure that the investigator gets a clear view of the credit transactions done by the customer and if there exists any abnormality.

6. KYC information

Ask any investigation officer and mention of KYC will always appear in their investigation process. Basic information like intent of setting up account, expected turnover, source of income, etc., are generally cross-referenced for any investigation. Getting it handy in the investigations screen will be an important aspect of the investigation.

7. Adverse media

One of the important aspects for any investigator is to search for adverse media news on the investigated customers. It is generally a time-taking and labor-intensive exercise. Basically, the investigator is trying to leverage additional sources to know more about the customer. The investigator is also trying to evaluate if there are a few developments that could be relevant for the investigator as a part of the investigation.

This could be a necessary albeit time-consuming manual activity for the investigator.

8. Customer network analysis

One of the most important parts of any investigation is the network of any individual. The type of transaction behavior that the individual is exhibiting. More details on the type of counterparties, type of products, and quantum of payments done or received; all this information becomes very valuable in deep investigation of a customer transaction behavior and giving explanation to the anomalies observed in the alerts.

If it is not aided by digitization, each investigation can take hours to build this profile and then investigate.

9. Narration for case investigation

Once the investigator has analyzed the transactions, he is supposed to close the alert with a disposition. Depending on whether the investigator found anomalies and sufficient reasons for feeling suspicious or vice versa, it needs explanation and summarizations of the findings. These narrations can be time-consuming for an investigator. A good functionality would be to leverage natural language processing to generate few automated insights into the form of narrations. It reduces the time for an investigator and helps in providing relevant insights before closing the alert or case and its outcome.

9.3 Application of Artificial Intelligence in Optimizing Investigation

Before we delve deeper into application of AI in this domain, let us first understand the challenges of an investigator. An investigator is looking for transaction patterns, proofs of credits or debits, ensuring that the customer being investigated is not in the news for wrong reasons, consolidating all information about the customer being investigated to provide an opinion and perspective on the customer's activity and classify them as suspicious or normal.

Ultimately, it is a judgment made by the investigator and is backed up by sufficient inputs from different sources that form the basis of taking such a call. During our interactions with multiple stakeholders in this domain, we have seen introduction of bots to do many of these activities. The biggest challenge in these cases has been the efficiency. If one does a work time and motion study (a study used to observe the activities done by workers on the shop floor, with an intent to identify inefficiencies and thereby improving productivity). It will become apparent that each of this information has three issues:

- Varying degree of automation
- Different capability to generate insight across each subject area
- Scattered pieces of bots and tools that maximize the time of an investigator or an analyst to compile them to generate big picture.

These issues are challenging enough for the investigator to go slow and not able to maximize the productivity that could be expected. It will also mean very high degree of concentration and scope of clerical errors. These further tire up the investigator. All of them would translate into further slower speed of analysis, higher fatigue, higher clerical errors, and overall low productivity.

Artificial intelligence needs to solve these specific problems to make life of an investigator easier and enhance the effectiveness and productivity (Fig. 9.2).

A few of the information and functionalities like triggering transactions, alert history, and customer consolidation of transactions is already provided by existing case management systems. These pieces of information can be easily consolidated and delivered to the investigator.

However, the following are the key opportunity areas.

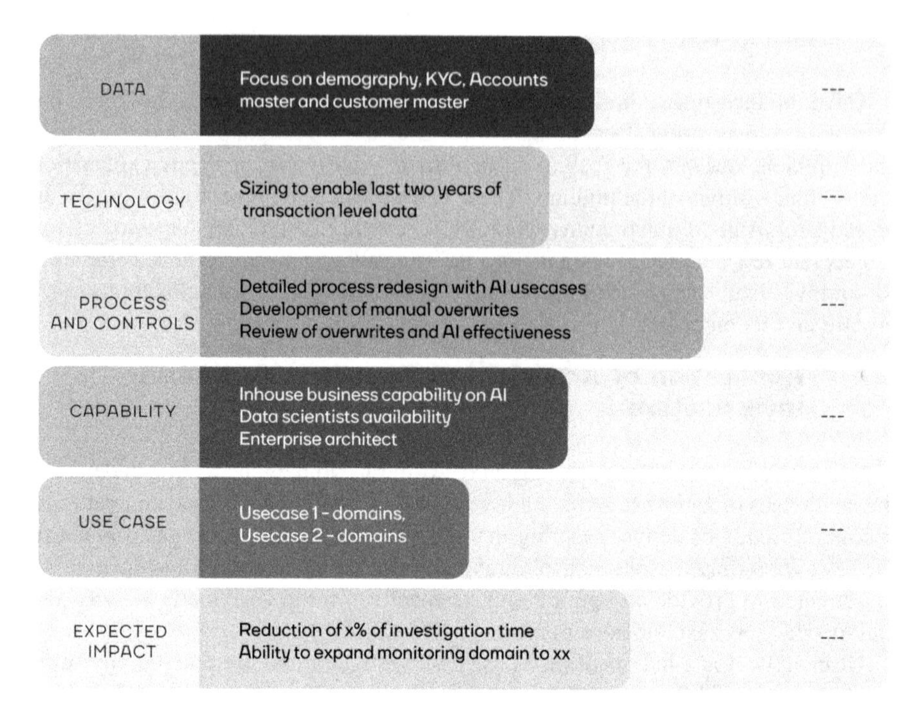

Fig. 9.2 The areas of opportunities are explained

9.3.1 Automation of Analyzing Transaction Behavior

Investigator spends a lot of time analyzing the transaction behavior of a customer. It requires the investigator to look at past transactions, summarize them in terms of trends and patterns and then analyzing, if they seem to create a bigger story in terms of insights and trends.

These insights can be easily incorporated into automated solutions. A quick mapping and automation of such solution will easily solve this manual activity that the investigator needs to do. It ensures that investigator already has all the calculations on the fingertips. They just need to study those numbers and generate inferences.

9.3.2 Network Analysis

Networks can be shown through visual representation to enable better tracking of transaction trails. It otherwise becomes difficult to understand the linkages—especially structuring. Networks can also be derived through social network analysis (SNA) variables. Some of the important ones are link functions, centrality, multiplexity, density, and so on. Through some of these functions, we can also contextualize the relative riskiness of the customer. Artificial intelligence becomes

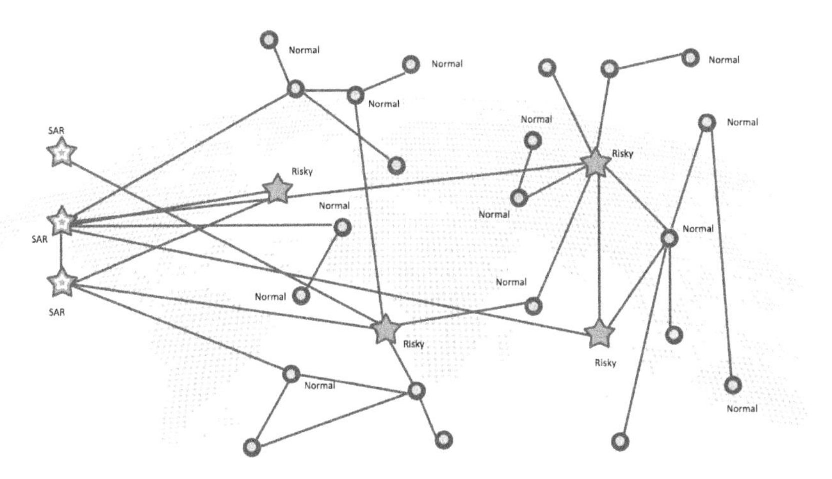

Fig. 9.3 Provides a conceptual understanding of the networks that any customer can have

a good way to extract these meaningful interactions and then summarizes them for the investigator.

A lot of insights about customer and the transaction behavior lies in the transacting parties. A regular dealing with a money launderer, structuring of money to route it to ultimate beneficiary, and many more of such transactions can be traced easily, if the network of the payments is well-organized and is available to the investigator visually (Fig. 9.3).

Another pointed input can be highlighting few networks that are seeming risky or suspicious.

9.3.3 Automation of Adverse Media Search

Imagine the benefit of all major relevant articles that are published already neatly organized for an investigator to see and know more about the customer under consideration. This can be achieved by leveraging unstructured text mining, a bit of web technologies, and text summarization analytics. Application of this can significantly reduce the investigation time. If an investigator wants to search more, it can always be done. However, for more than 50% of the investigated cases, there exists an opportunity to reduce the time spent on this activity.

9.3.4 Automation of Narration for Case or Alert Closure

Depending on the type of transaction behaviors, customers will exhibit varying patterns and trends. It would be a big help for technology to automatically view those themes of trends and patterns, identify anomalies, and document them in the form of narration.

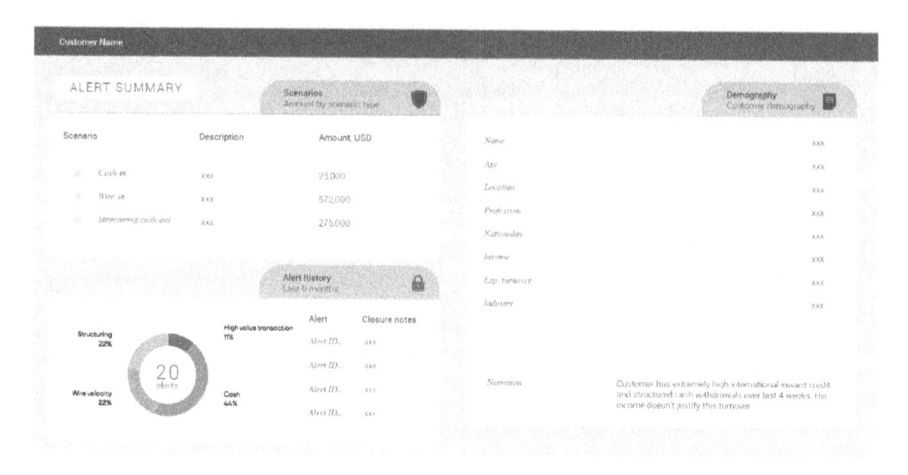

Fig. 9.4 Showing single investigation view for the investigator

Application of natural language generation can bridge this gap. This helps in significant reduction of investigation time, especially for low-risk alerts as investigator doesn't need to type anything. It will only require reviewing and modifying the text and closing the alert.

In case of high-risk alerts, investigator has already got summary of all the key observations. It becomes very easy for the investigator to review, update, and close the narration.

This becomes a significant improvement in productivity for the investigator (Figs. 9.4, 9.5 and 9.6).

These exhibits are indicative. Eventually, organizations will have to customize the investigation screens, based on their own needs. What is worth highlighting is that the design will be adjusted to what their investigators analyze. Create them upfront and convert them into visual representation. This will save time as well as give necessary insights upfront.

As we mentioned above, depending on the sophistication, multiple financial institutions are leveraging various aspects of automation or at least providing this information to their investigators. What is missing among this is the consolidation and minimal diversion to collect and collate this information. A good solution should be able to provide what we would like to call "a 360-degree investigation view" on a single click and in an automated manner to reduce the investigation time as much as possible.

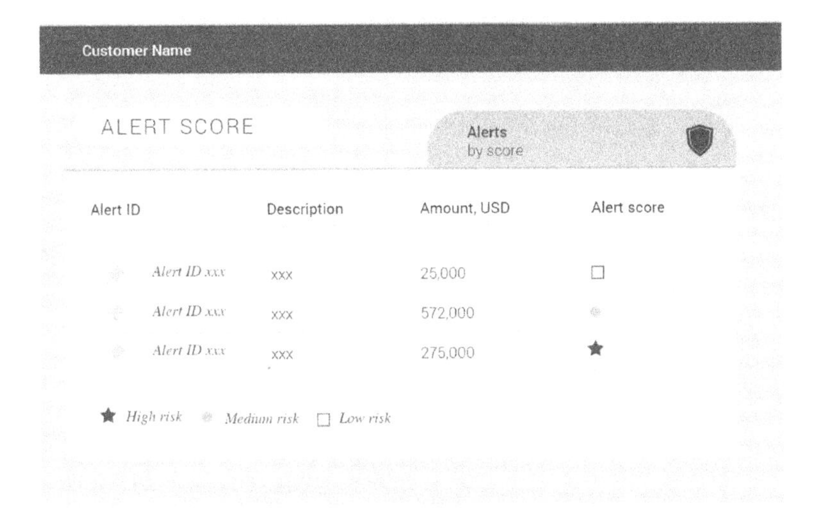

Fig. 9.5 Showing single investigation view for the investigator—continued

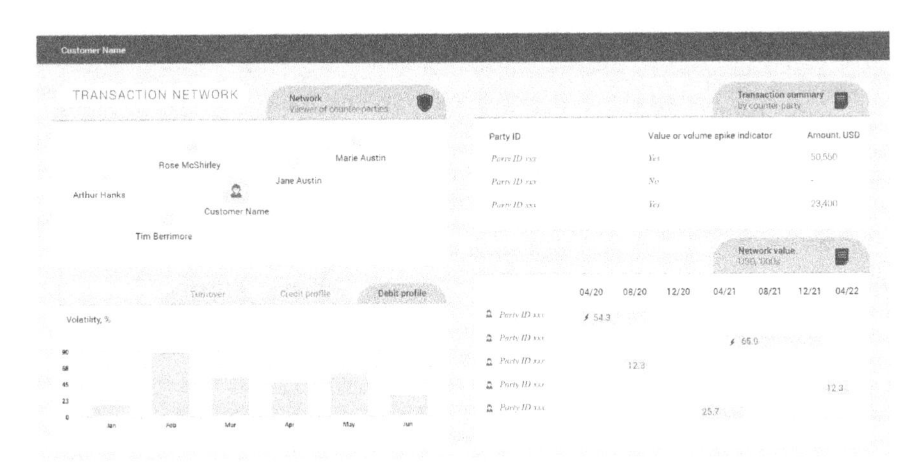

Fig. 9.6 Showing single investigation view for the investigator—continued

10 Ethical Challenges for AI-Based Applications

10.1 Introduction

Ethical AI has been in practice in some parts of the developed world for years on topics like credit decisions and customer acceptance for doing business where it is clearly mentioned that key discriminators for rejecting an application should be transparent. While race, gender, etc., cannot be used to discriminate against customers.

Application of AI getting it wrong is widespread. Amazon realized that their AI-enabled tool has a bias against women. My wife enters India, and she gets a lot of advertisement on ethnic jewelry and sarees from Instagram and when she enters Canada, first thing Instagram pushes to her is surrogacy advertisement.

While ethical AI has been a topic being discussed by practitioners for a long time, one unanimous point that came out from various interviews is that conversion of practices of ethical AI in day-to-day data sciences is not in widespread practice.

Another point mentioned by a few practitioners was that in compliance there are typologies where people from specific nationalities are profiled as high risk. Historically, profiling, in general, has been demographically driven. Would that be seen as bias against certain groups?

Then there are issues related to customer privacy. Processing unstructured data and mapping network information go quite deep into some of the information that could technically be considered private information and would be governed by consumer privacy.

So far there are no reported incidences, but there exists a business case of financial institutions using AI to discriminate against customers.

While the benefits of leveraging AI are evident in the realm of financial crimes, they are not without their risks. To prevent the misuse of customer data and personal information, banks must develop an ethical framework and validate the decision-making processes behind algorithm applications.

A. Gupta et al., *Artificial Intelligence Applications in Banking and Financial Services*, Future of Business and Finance, https://doi.org/10.1007/978-981-99-2571-1_10

Let us first understand the issues underlying the ethical AI, then the ways to handle or mitigate the risk would be assessed.

10.2 Misuse of Personal Information

One of the commonly raised and debated concerns is that AI could lead to the misuse of personal information. While in general the benefits far outstrip the concerns, there exists a clear downside of personal information usage.

This is coupled with the varying standards of privacy for the collection of data and privacy protection for sensitive financial information. The current legal framework for privacy and anti-discrimination measures is inconsistent across jurisdictions and industries.

The European Union has already implemented data privacy laws. Regulatory landscapes are expected to move toward a more European model. These laws require that banks restrict their use and access to personal data. However, the downside of consumer privacy in their initiative to create an ultimate beneficiary record is also becoming evident. Hence, it is imperative for the regulators to understand and appreciate genuine needs of the compliance department in knowing the information. Create checks and balances around misuse to ensure that while data privacy is not compromised, it still helps the business in getting necessary insights for aiding decision-making.

Financial services companies, whether regulated or not by their local regulators, need to be mindful of these privacy concerns and should proactively develop governance framework for dealing with such issues.

10.3 Introducing Bias

Bias can be introduced into AI from various sources. Our discussions and research have pointed to three specific types of biases so far (Fig. 10.1).

Human bias:

It refers to the biases that humans carry from their past experiences. It could be conscious or subconscious. As they are mentioned, they could be conscious biases that are generally observed even through interviews. There would be clear cases of certain ethnicities associated with a specific type of behavior in the mind of decision-makers. Part of these is also backed by the data. There are instances when the decision-makers pointed out a group having relatively higher risk behavior, or the group has a relatively higher incidence of laundering money as compared to others. That insight from a relatively smaller sample stays with those decision-makers forever. This can bring biasness in the design of experiments, selection of data, the framework of modeling, and so on.

It can have multiple implications like preventing a FI from meeting their diversity targets, improving community outcomes, and creating more diverse and

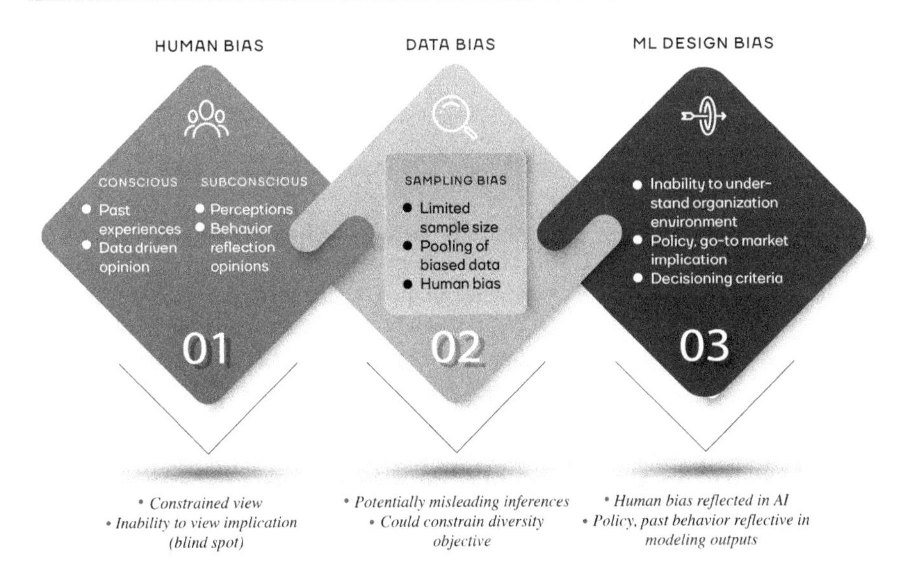

Fig. 10.1 Showing different types of biases in the development of machine learning models

inclusive environments. Different neighborhoods, different communities could be adversely targeted.

Data-driven bias:

This bias is introduced due to various reasons—poorly classified or sampled data being the major cause. There are so many poorly conceived sampling-based decision-makings that we as consultants have come across in various organizations.

Biased AI algorithms may cause unintended consequences, such as the exaggerated treatment of bias against women in Apple's credit card application. Sometimes pooling of data from different sources can also introduce bias. While gender-based discrimination is prohibited in fair lending laws, it is allowed in auto insurance. If we pool the data from these two, we would unknowingly be introducing bias in our data.

In one of the organizations, the client had very little data for a segment, where they wanted to create a model. They were advised by another consultant to enhance the data through simulations. Interestingly, the consultant enhancing the data did not try to understand underlying population distributions, behavioral distributions, or segments that they represent. They just expanded the sample size by using simulation—essentially increased the sample size, with a mix of underlying distribution of limited observed event cases.

While it gave them a bigger sample size, the concern of lower sample size remains. The organization still did not have a good and stable view on the population that they were considering high risk. It was just replicating limited view to more rows.

Another instance was of poor-quality missing data from a few cities for a model development exercise. Challenge in these cases is that missing values are not missing at random. Unfortunately, these get clubbed with other missing, who can have a very different risk behavior.

These biases can impact the model and the treatment AI can give to a part of the population.

A publicly known case of bias was Microsoft's chatbot, which exhibited racial and other biased sentiments. It was trained on tweets and was taken off the market after the company realized it. Removing demographic information isn't enough to prevent bias, since these data are necessary for determining the bot's abilities. These biases come from the data used, social contexts, and the way data is represented.

10.4 Bias in the Design of Machine Learning Algorithm

We discussed modeling framework in detail in Chap. 8. It dealt with different aspects of model design, definition of sample, the definition of events, and other predictors. This could be another area where for example a policy from the institution would result in certain behaviors or segments available and certain others not available. This can lead to disproportionate information for the machine learning and hence slightly biased results. For example, AI can't detect biases in borrowers for a mortgage portfolio. It means that if the algorithm decides based on biases in data, it is likely to repeat discriminatory mortgage lending practices.

Mimicking a real-life scenario, ensuring that filters put in data are relevant. Knowing the segments underrepresented and creating a treatment for them or at least limited pilots to collate data to train the behavior would be some of the important considerations to be pursued in a real-life model situation.

10.5 Ethical Challenges in Developing AI for AML

The challenge with biased AI is ensuring that the model is free of bias. The AI community has discussed this issue extensively. In the AML program, this is most evident in the know your customer (KYC) area. In this area, the data science team must ensure that they use features that do not encode systematic bias. This is not the case in the transaction monitoring area, which deals with transactional data. AML systems could be biased. In some cases, the system's decisions impact a certain group disproportionately. For this reason, banks should carefully consider the ethical implications of the technology they choose. For example, an AI system may not be unbiased if its choices negatively affect a certain group of people. If the system is biased, the bank should adjust its policies.

Currently, most of the AI used in banking institutions is limited to low-level decision trees and robotic process automation. More sophisticated applications of AI are being developed, which can pose ethical issues. The banking industry must also ensure that these advanced technologies are used with appropriate care, ensuring fairness and security of consumer data. Ultimately, ethical challenges are essential for the success of these systems. They can even affect customer trust, which is crucial for the stability of financial markets.

10.6 Ways to Avoid Bias in AI

While this is a relatively newer area, it is getting enriched with continued usage and research. There are few best practices. Some of them are used by us modelers for quite a few years as best practices. Some are introduced recently with our evolving understanding of the topic and better appreciation to the downside of such practices:

Data:

Ensure that the sampling methodology is fairly robust. If required, insist on explicit definition of segments and then follow stratified random sampling. If alternate techniques are used, then focus on the representation. As a modeler, we always use to focus on the representation of the data. It is important to conduct representation analysis across some of the key segments.

Validation of model:

Ensure that the model is predictive across segments. We generally conduct segment-wise predictive power whichever is used—GINI, KS, precision. However, it has to be done for segments. Even if the team accepts the recommendations, they need to understand and acknowledge that model would not work very well for a few segments and there has to be plan before the model rollout.

Decisioning analytics:

Simulate the decisions on key segments based on model outcome. The results could be a good litmus test which combine impact of policy, model, cut-offs and overlaying strategy in decision-making.

It would be easy to generate results like acceptance, impacted segments, impact on company's financials, and intended target groups. It would give a clear view to the senior management if there exists a bias and if the bias is behavioral or targeting a demography. Senior management can still override few decisions, if their strategy is clearly to mitigate impact of some of the learning to a few targets group.

Organizational initiatives:

Organizations can create multiple initiatives outside of technical realm of bias introduction for example third-party regular review of biasness in AI, sensitization programs for the AI professionals around what constitutes bias and how can that be managed. It could even be extended to diversification in their AI and data scientist teams that will bring a cultural change and a diversified view on the same data and models through different lens.

There are few open-source resources like Google has provided a resource titled "Google Responsible AI practices" that provides a good view on ways to introduce fairness in AI. IBM has introduced a whole library on GitHub that allows users to conduct bias tests. The library is titled "AI Fairness 360".

Most of the new models Ops platforms like DataIKU, SAS, and IBM also have resources that focus on model explainability and assess biasness to some degree in an automated environment.

To summarize, while this new field is slowly providing impetus to fair AI that has limited bias, it will be a long journey for all AI practitioners as much as AI in compliance.

Setting up a Best-In-Class AI-Driven Financial Crime Control Unit (FCCU) 11

11.1 Introduction

Throughout our whole book, we discussed various aspects of an intelligent and efficient compliance organization. Each of the topics that are discussed has explored different aspects of processes that help in optimizing the compliance function.

Organizations don't make a big bang self-transformation approach. It becomes too disruptive for everyone. Every organization must do a self-assessment and then creates a road map for itself. However, one common theme that exists across organizations is that they traverse a journey. Better-managed ones, start with the end state in mind along with the milestones. Not so well-managed one takes things on a piecemeal basis. Eventually are lost in use case implementation approach. Irrespective, it helps the organizations take a leap and upgrade themselves to a better and more efficient compliance organization.

In this chapter, we would define our version of a best-in-class compliance organization. Our best-practice organization has a few important aspects.

A best-in-class financial crime investigation unit is a cross-functional team that has experience of AML experts, data scientists, financial analysts for corporate clients and if possible legal or law enforcement specialists who bring broader experience in dealing with such cases.

The objective of such a team would be to focus on generating high-quality alerts that are likely to be money laundering events. This is done by using sophisticated data analytics and combining internal and external data sources including negative media, and other external sources. The other aspect of such an organization is a technology-enabled sound investigation system that prioritizes cases based on risk assessment. It also entails a well-trained team that is proactive in identifying hidden patterns of transactions and reducing false alerts while providing feedback on emerging trends for continuous learning of machine learning and data analytics.

A. Gupta et al., *Artificial Intelligence Applications in Banking and Financial Services*, Future of Business and Finance, https://doi.org/10.1007/978-981-99-2571-1_11

Advanced analytics and data mining are transforming the way the financial services industry investigates and fights financial crime. This data helps investigators identify money laundering or terrorist financing transactions and other criminal activities while eliminating false positives and enhancing investigative processes. The aim of leveraging technology and intelligence in such an organization will be to maximize efficiency and increase surveillance while continuously reducing and optimizing operational costs.

These units combine artificial intelligence, behavior analytics, data insights from structured, unstructured, internal, and external data to help organizations proactively detect, investigate, and report such crimes.

To achieve this, there are a few key organizational drivers. Let us discuss them:

(i) Leadership
(ii) Organization design
(iii) Team capability
(iv) Data organization
(v) Technology setup
(vi) Machine learning management
(vii) Productization of business ideas.

11.1.1 Leadership

Leadership for an organization plays the most important role in defining an organization. The culture, the people, the capabilities, all reflect the kind of leadership an organization possess. For a compliance organization the leadership of relevance is at two levels—the chief executive officer (CEO) and the chief compliance officer (CCO).

While the board of directors provides overall vision and direction to the organization, it is translated into the tangible path to pursue, by the CEO. A well-functioning compliance organization has a hands-on CEO, who along with CCO ensures that all senior stakeholders are aligned with the end state of the compliance organization. CEO also enables knowledge enhancement sessions for the senior management and the BOD to ensure that when the relevant topics are discussed, all stakeholders have a fair understanding of the best practices, regulatory environment, and the outcome which is expected from a well-functioning organization. It enables knowledgeable discussions, active tracking of progress, and alignment on rewards for the performers, who have delivered as per the roadmap. CEO himself takes a lead in regular reviews on the strategic direction and milestones achieved rather than engaging only in operational updates. The operational updates, firefighting, and response to audit comments from the central bank are critical as they ensure smooth functioning of the organization, but also it is important to keep a tab on where the compliance organization is headed and when can minor operational issues stop becoming daily affair by applying intelligent technology to centralize, make organization efficient and ensure controls are in place. A good indicator for a CEO would be to assess market environment internationally, benchmark itself with the best of the organizations, ensure milestones are more often met, and the number of operational issues, internal audit observations, and human resource costs, as well as technology cost, is getting optimized.

Chief compliance officer (CCO) is the actual executioner and baton holder for the organization. CCO is not only well-educated about key themes around compliance globally, CCO is also made aware and is appreciative of artificial intelligence, but the role machine learning also plays in optimizing them. CCO should have an advisor who is knowledgeable in machine learning and digitization-related topics. The reason we are highlighting this is that majority of the business stakeholders in compliance have traditionally been in non-data and non-AI backgrounds. Many a time, they understand that there exists an algorithm that can help businesses do better. However, they lack an understanding of the nitty–gritty of algorithms, and how they are developed and implemented. They sometimes also lack understanding of the reasons for poor-quality models and hence can't provide executive sponsorships on data governance or analytics programs, thereby leaving these domains waiting to be improved and made best in class. This remains a gap despite large investments in technology and data platforms.

A sound advisor can bridge this gap for a CCO and can provide perspective on technical topics while a CCO focus on organizational capabilities, processes, people, and technology alignment.

With regard to KPIs, CEO and CCO will share the majority of the KPIs. Just that CCO will have more direct oversight and control on various drivers of the compliance organization. CCO will also have direct responsibility for the performance of the team and their capabilities as well as delivery. This way, the vision, and alignment can be ensured starting from the top to the bottom rung of the organization.

11.1.2 Organization Design

Another important aspect of an organization is the organization design for such an organization. While every organizational design has a context. The topic needs sufficient deliberations among the stakeholders and then decide on the structure. However, we are defining rather a generic organization design for such an organization. This can be customized to the organization depending on the constraints and responsibilities that the compliance department needs to assess (Fig. 11.1).

As shown above, there are five departments that we would recommend. The size of these departments can vary and depend on the size of the organization, some of these functions can roll into the same individual or same role. The various departments of the organization are as follows:

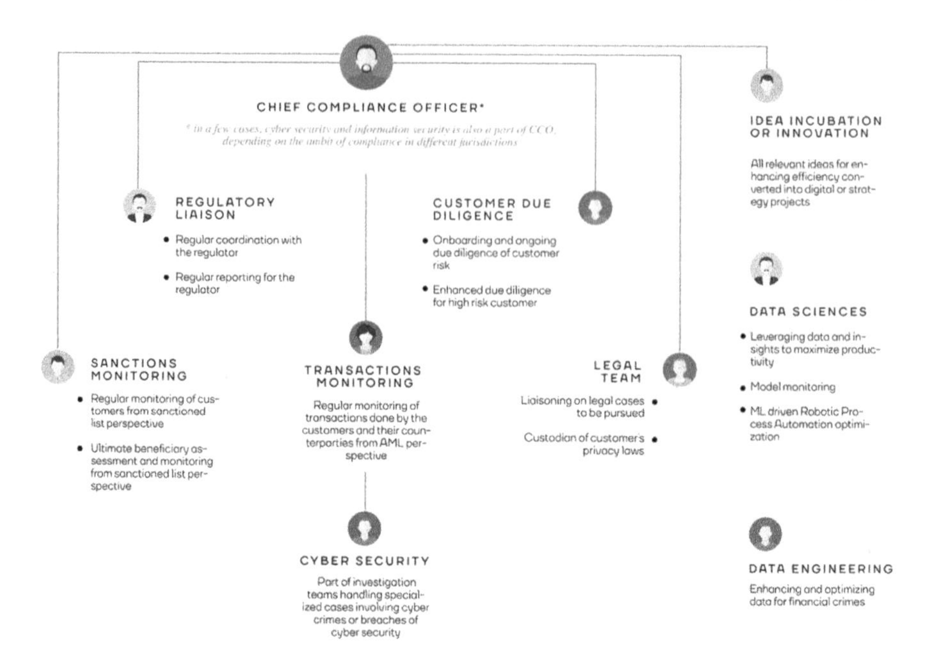

Fig. 11.1 Showing an organization chart for a well-performing AML and CFT department

Solutions team—This is typically a part of the IT organization. The responsibility for this team is to keep the solution functional. All infrastructure management issues, solution functioning, data loading and software maintenance, and patch or hotfix applications are the responsibility of the solutions team.

Investigation team—team's responsibility is to investigate and close alerts that are generated. It could have subdivisions for financial transactions, onboarding, and real-time sanctions screening.

Regulatory team—this team has the primary responsibility of dealing with the regulators. There are two specific functions that the team performs—(a) coordinate with the legal team and report the incidence of a suspicious activity or watchlist matches; (b) coordinate with regulators for ongoing audit and coordination as well as reporting.

Analytics team—the responsibility of this team is to generate insights for the compliance organization, develop machine learning-driven solutions, work on the adoption and customization of AI applications, and provide business intelligence support to senior stakeholders.

Another important responsibility of this team is to maintain the model lifecycle and develop champion challenger strategies and models. Measure and report the model performance and make business cases for introducing and phasing out the analytical backbone of various tools that are leveraged by the compliance teams.

Data management team—this team like the infrastructure team, can be lent by the group data management team or it can be a standalone team. The prime responsibility of the team is to ensure data governance and improving overall data quality within the organization. Ensuring seamless and secured data access to the relevant stakeholder is another responsibility for the team. This team will also be responsible for implementing GDPR or data privacy regulations within the organization.

Cybersecurity team—this is a relatively new team. One of the reasons for the addition is that the money laundering, terrorism financing, fraud have evolved to liberally use IT-related devices and networks. Other challenge that financial institutions face is that their own system gets prone to information security issues. It is impractical for untrained resources to investigate such crimes. Trained staff would use information like Internet Protocol (IP) addresses that can help locate the specific location of the perpetrator or details on the device from which the logins are made and any such patterns can also be a part of the investigation. Given these developments, cybersecurity has become an integral part of the financial crime management program.

For multinational organizations, most of these teams are replicated in a scaled-down version. As most of the regulators have barred the data to leave the physical borders of the country. This means that solutions, data management, and investigation teams are necessarily replicated. However, the analytics team could be a centralized capability. Ensuring best practices, coordinating on quality of models, conducting regular validations, etc., must be centrally monitored activities. Similarly, data management can be locally handled either as a part of an enterprise data management team of a particular country. It could also be self-sustained data management team for compliance, coordinating with enterprise data management team depending on the size. The control of data quality and usage of data for intelligence still must be centrally controlled. It ensures that organization can navigate the critical path items with a clear vision of the headquarter (HQ). HQ also has visibility and control over critical activities which are shaping up the future of compliance functions in individual countries. The rest of the functions will also have their counterparts which are generally handled through internal audits, enterprise-wide IT policy, and similar policies or audits.

Another option can be that multiple functions from solutions to investigations are outsourced to managed services. In that case, it is imperative that the analytics team still work closely with the managed service provider, as the inefficiency in alerts, or watchlist or investigations will be eventually billed to the financial institution. In this case, the knowledge-driven services could still be centralized and then coordinated with managed services provider to optimize the whole process.

11.1.3 Team Capability

Over and above regular job descriptions and their roles and responsibilities like regulatory coordination, legal, IT, and investigation operations; there are a few critical capabilities that should be added and validated for a futuristic AI-enabled organization:

Advisor to the CCO—CCOs are increasingly getting exposed to data, machine learning, and emerging technologies. They ideally need an advisor, who understands these emerging themes, capabilities and can demystify them for the CCO. It need not be a fulltime role. It can either be taken from outside the organization on a need basis, alternately, if the organization has an in-house center of competence, then the skillset can also be leveraged from there. Key to this role is a clear definition of responsibility, KPI, and performance evaluation. The advisor cannot have free lunches in this role.

Resident data scientist—A resident data scientist, who is a part of the analytics team. Whether the team develops the models in-house or they outsource them to external parties, the ownership of the models, their adoption, the performance monitoring, and ongoing evaluation are critically important. These must be managed in-house through the data scientist. Data scientists can also evaluate emerging use cases and validate the analytical approach along with thinking through the pitfalls if of such an approach. Their KPI must be aligned with the business outcome. It means models once developed must be adopted by the business, and the impact has to be assessed and improved on an ongoing basis.

Resident data engineer—this capability will reside within the data management team. Data engineers are entrusted with ensuring that structured and unstructured data pipelines are established. They are being managed efficiently. ETL processes are executed properly with regular business validations. Lastly, the ongoing data quality improvement projects or leveraging unstructured data-related projects are pursued under the guidance of the data engineer.

Enterprise architect—readers will notice in the machine learning section that we will propose increasing usage and collaboration from Fintech organizations. While Fintech accelerates the innovation and adoption of the solution inside the organization, it is organization responsibility to ensure that the solution is seamlessly adopted and integrated into its IT architecture. This is precisely the role of an enterprise architect. Again, this can be borrowed from Enterprise IT team; however, this capability must be managed by engaging all stakeholders, responsible for executing the transformation strategy.

11.1.4 Data Organization

Data plays a significant role in futuristic organizations. There is an old saying—"garbage in garbage out". Despite, it being quoted so often, very few organizations really go top-down in managing the data organization. In this chapter, we will not discuss data governance, which has already been discussed by many authors. However, translating implications of data infrastructure from a compliance perspective, developing roadmap on data infrastructure, managing quality of data, maintaining security, and accessibility of data are few of the topics that warrant in-house ownership of these. KPI for the data organization lead is to present the milestone achieved, that are commonly set by the organization in terms of the data journey. More importantly, the performance needs to be validated by the business users and data scientists to ensure that milestones are indeed delivering desired results.

11.1.5 Technology Set Up

Most of the solutions have a legacy system for managing, monitoring, and filing regulatory reports on AML- and CFT-related functions. With the increasing complexity and dynamic evolution of the regulatory framework, financial institutions are increasingly getting challenged to cope with the fulfillment of the new requirement through existing solutions.

Investment in new solutions is a double-edged sword. Organizations take years to introduce and stabilize a system. Hence, it is always tricky to upgrade and migrate to different solutions. While the discussion on IT systems is done in detail in the previous chapter, here we would highlight the solution capabilities that exist for a well-performing organization:

1. Solution is flexible enough to customize the scenarios. It would be better if a solution can also accommodate machine learning algorithms to manage different types of AML monitoring
2. Solution has a flexible architecture to make changes on the investigation screens. It ensures new dimensionalities of channels and trends can be easily configured and managed
3. The solution should have the capability to execute machine learning algorithms and data mining algorithms
4. The solution can have the capability of robotic process automation (RPA) to automate and integrate all possible investigations through a common interface.

Organizations will need to take a call on their solution continuation. Points three and four above can be plugged into existing solutions. However, if the first two points are not easy to manage, it is time, that organizations should think about migrating to upgraded platforms.

Another important aspect of the organization is the agility to develop and adapt to the new features and functionalities. Our experience suggests that large organizations struggle with innovation. It is in this area, where CCOs should direct their teams to engage with fintech, and regtechs and can pose specific business problems for them to solve. FinTech is also agile enough to bring and integrate the solution into their legacy system without much hassle. This to us would be the way forward, i.e., enhance capability of your core products by collaboration and eventually migrate to a more capable product.

Lastly, the solutions are migrating to the cloud. Concepts like DevOps for solution development and model ops for model management are increasingly becoming common. Creating local GitHub and creating clear policies and mechanisms for model inventory management and upgrade has to be wedded to machine learning management.

11.1.6 Machine Learning Management

The whole pretext for this book is to develop and manage an intelligent AI solution that optimizes the operations of compliance. It implies that machine learning models' lifecycle must be managed effectively. Starting by identifying new areas for applying machine learning models, to the development of such models, then deploying, monitoring the performance, and replacing those with more efficient or more applicable models is an ongoing exercise. Model governance is an important topic that needs sufficient attention. A well-functioning organization proactively understand and manage the model governance (Fig. 11.2).

This summarizes the structure and capabilities of a best-in-class organization. However, when we were discussing the end state and the roadmap that is to be developed. The roadmap will primarily comprise the above six elements that should be benchmarked with a clear sense of where the organization should be on each of those six elements. Eventually, every organization will define its end state. The end state need not be the perfect world benchmark if the organization is clear about the reasons for defining a particular end state.

This will ensure that the organizations will have a clear sense of their journey and where they must reach, rather than piecemeal approach.

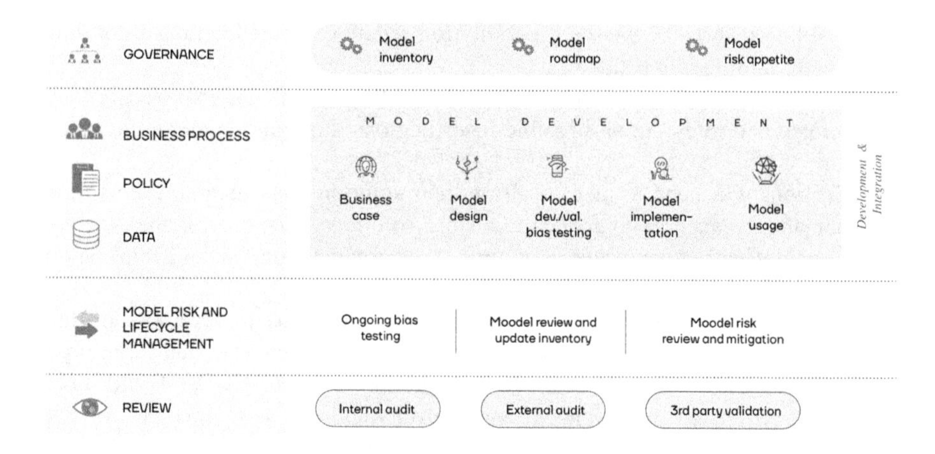

Fig. 11.2 Provides overview of model governance

Appendix 1: This Section Focuses on the Evaluation of Various Machine Learning Techniques

There are traditional modeling techniques like nonlinear regression class techniques—the most prominent among them is logistic regression. There are classification algorithms like decision trees and random forest. There is also the option of structured equation model.

Another category of models are ensembles which sequentially train the model with the gradient boosting method. The name of the technique is called XGBoost.

There are also a set of methods available among deep learning like LSTN, RNN, etc. within categories of neural networks. Each of them has its own applications, depending on the type of problem one wants to solve.

There are also questions that most practitioners pose to data scientists is "how do you select a modeling technique" and "how do you decide that the model developed is good. The efforts to improve the model should stop?".

This appendix is broadly divided into two sections:

1. Selection of the right technique for training a machine learning model
2. Finer aspects of finetuning and finalizing a model.

1. Selection of the right technique for training a machine learning model

For the current discussion, we have kept the focus on predictive modeling. The reason for keeping this discussion specific to predictive modeling is that majority of the business problems around financial transaction monitoring are converted into predictive model problem. While majority of the practitioners focus on machine learning problems related to financial transactions being predictive models, a modeler can also convert it into a segmentation problem. We would not get into an academic debate on which one works, we would rather focus on what is used more often.

Once the prediction on the outcome is done, then based on the cutoff, the customer or the transaction is classified.

We will do a short introduction to the important modeling techniques. While each of them requires a separate technical and business paper, our objective is not to provide a technical explanation of these. We would do a quick introduction of

A. Gupta et al., *Artificial Intelligence Applications in Banking and Financial Services*, Future of Business and Finance, https://doi.org/10.1007/978-981-99-2571-1

these techniques before our assessment of a few empirical comparison on their effectiveness.

Logistic regression

Logistic regression is a specific type of a generalized linear model (GLM) that uses logistic function to model dependent and independent variables. The model is developed on binary outcomes by calculating their log odds ratio. The model is calibrated by looking at predictors (independent variables) against the log odds of the outcome (event). These log odds can be converted into a probability to infer the output. While there have been many advancements in the machine learning techniques and there are better techniques available, it is still one of the most used techniques. One of the major benefits of using logistic regression is its simplicity. Both in terms of developing the model and the model explainability, logistic regression models are easier to train and set up than other machine learning algorithms. Another benefit of logistic regression is its robustness to increasing event rate. Logistic regression models have been seen to provide more stable area under the curve (AUC) as the number of events is increased in the model, as compared to other techniques.

Decision trees

Decision trees are a set of models that are classified under objective segmentation. Under this method, there is an objective function or the event (financial crime for example as 1/0). The tree is grown using different techniques depending on the underlying data type of the prediction event. Most commonly used technique for binary outcome or multinomial outcome (0/1/2) is Chi Square Automatic Interaction Detector (CHAID).

In this technique, a tree is grown in the form of nodes. Each node is a variable used as an input. CHAID ensures the ordering of nodes in the descending order of the chi square value to ensure that relatively higher predictive variable is placed as a higher node than relatively lower predictive variable.

The benefits of decision trees are a few: (a) they are quite intuitive to infer. As the combination of nodes can translate into a segment, this becomes very easy for the business users to understand that which segment is driving differential behavior and which one does not, (b) decision trees work on multinomial data easily. If you have multiple outcome (0/1/2) while logistic regression can also give output through multinomial logistic regression, but the precision of right classification is relatively poor, and (c) majority of the solutions that leverage decision trees give user an option to force fit a variable based on their business judgment and can still grow the tree.

Random forest

Random forest is an ensemble technique, which overcomes a lot of downsides of decision tree like overfitting, treatment of missing values, etc.

It is developed using multiple decision trees. The algorithm leverages a training process called bagging. Bagging basically entails selection of datapoints with replacement for the training of multiple decision trees. The outcome is the average of multiple trees, thereby making the process more stable and precise.

It is logical that increasing the number of trees will make the outcomes more stable.

Some of the benefits of the random forest are relatively better predictive models due to bagging algorithm. Like decision trees, it can handle large volume of data. This has been the next most popular technique being used in the last couple of years.

XGBoost

This is another ensembling technique that is currently used as one of the most advanced techniques for low to medium available data on event rates. Unlike bagging, this ensembling is sequential. It means that the error terms from model 0 are carry forwarded in model 1 and so on. This also helps in sequentially reducing the error terms to the best extent possible.

This technique is a modification to the traditional gradient boosting. The modifications are done through the introduction of gamma, theta, and lambda. These parameters provide a lot of flexibility to the modeler in terms of pruning of the tree, regularization parameter, managing effect of outliers, and speed of convergence. It is quite an interesting technique to do more technical reading.

In the recent experiments and literatures, XGBoost seems to be competing with random forest and neural networks, depending on the situation.

Deep learning models

Deep learning models are not a single model. They refer to a series of modeling techniques. Often referred to as neural networks. In general, the conceptual framework of deep learning models is to learn the underlying structure of the phenomenon, rather than explicitly modeling the trends in terms of hardcoded variables.

For example, a sequence of actions can be either modeled as variables that explicitly capture this sequence through derived variables. Another alternate mechanism for this could be learning these sequences in different neurons and layers. Then train these neurons and layers through sequence called forward propagation or backward propagation. The idea would still be to train the neurons to minimize the error through gradient descent.

Interestingly, the concept of neural networks is not new. It has been there since 80s. The only challenge then was the computing power and processors' capability. With computing power rising, deep learning has gained prominence.

It is one of the most prominent methods for training underlying complex interactions and patterns. While inference of models is not easy, the precision in unstructured training environment is a lot better, provided that there is sufficient tagged input data provided to the model.

Selection of right modeling technique

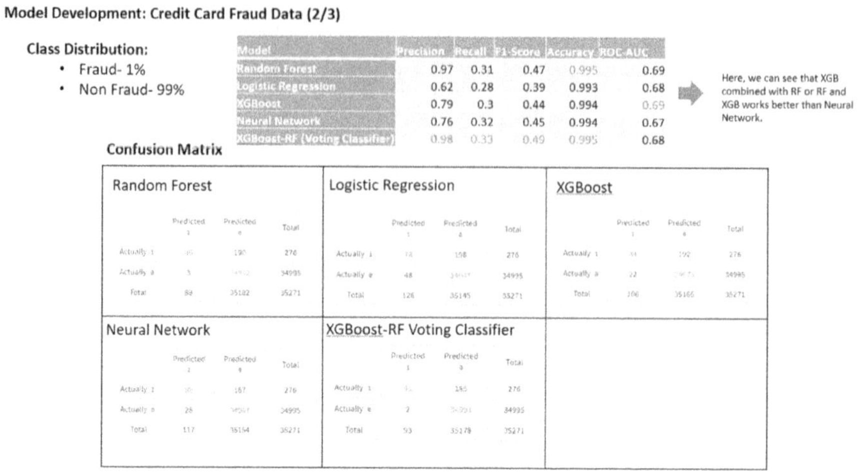

Model Development: Credit Card Fraud Data (2/3)

Class Distribution:

- Fraud- 1%
- Non Fraud- 99%

Model	Precision	Recall	F1-Score	Accuracy	ROC-AUC
Random Forest	0.97	0.31	0.47	0.995	0.69
Logistic Regression	0.62	0.28	0.39	0.993	0.68
XGBoost	0.79	0.3	0.44	0.994	0.69
Neural Network	0.76	0.32	0.45	0.994	0.67
XGBoost-RF (Voting Classifier)	0.98	0.33	0.49	0.995	0.68

Here, we can see that XGB combined with RF or RF and XGB works better than Neural Network.

Confusion Matrix

Random Forest

	Predicted 1	Predicted 0	Total
Actually 1	[illegible]	190	276
Actually 0	3	[illegible]	34995
Total	89	35182	35271

Logistic Regression

	Predicted 1	Predicted 0	Total
Actually 1	[illegible]	198	276
Actually 0	48	[illegible]	34995
Total	126	35145	35271

XGBoost

	Predicted 1	Predicted 0	Total
Actually 1	[illegible]	192	276
Actually 0	22	[illegible]	34995
Total	106	35165	35271

Neural Network

	Predicted 1	Predicted 0	Total
Actually 1	[illegible]	167	276
Actually 0	28	[illegible]	34995
Total	117	35154	35271

XGBoost-RF Voting Classifier

	Predicted 1	Predicted 0	Total
Actually 1	[illegible]	185	276
Actually 0	2	[illegible]	34995
Total	93	35178	35271

Fig. 1 Low-frequency event credit card fraud data

There exists a lot of literature that compares the techniques for their specific business purposes. Each of them would pick their specific business purposes and then apply the models to compare them.

To us, the consideration was slightly different. We looked at event rate (number of observations in the dependent variable) to compare the results.

We realized that one of the biggest drivers of varying performance of modeling technique is the event rate along with the underlying variability in the trends or patterns that needs to be modeled. While the latter is difficult to generalize, we looked at the efficacy of modeling techniques with varying event rates. For this, we extracted sample data from Kaggle. It had different event rates based on the type of model.

For example, a typical AML event rate would be no more than a percent, while high risk customer type of event rate would be higher. Ask any experienced modeler, the response of the modeling technique, and the modeling framework would change depending on the event rate that one mentions.

The outcome of the exercise is as shown in Fig. 1. When the event rates are lower or the absolute number of events are lower, then XGBoost would perform well. Random forest is the other technique which is recommended in that situation.

This is another example with relatively higher number of event rate. As shown in Fig. 2, once the event rate increases, XGBoost and random forest still give the best results, but neural network already started catching up in terms of model efficiency.

We have seen deep learning giving better performance with more unstructured inputs into the model. However, from Fig. 3, it suggests that deep learning models need not always provide the best results.

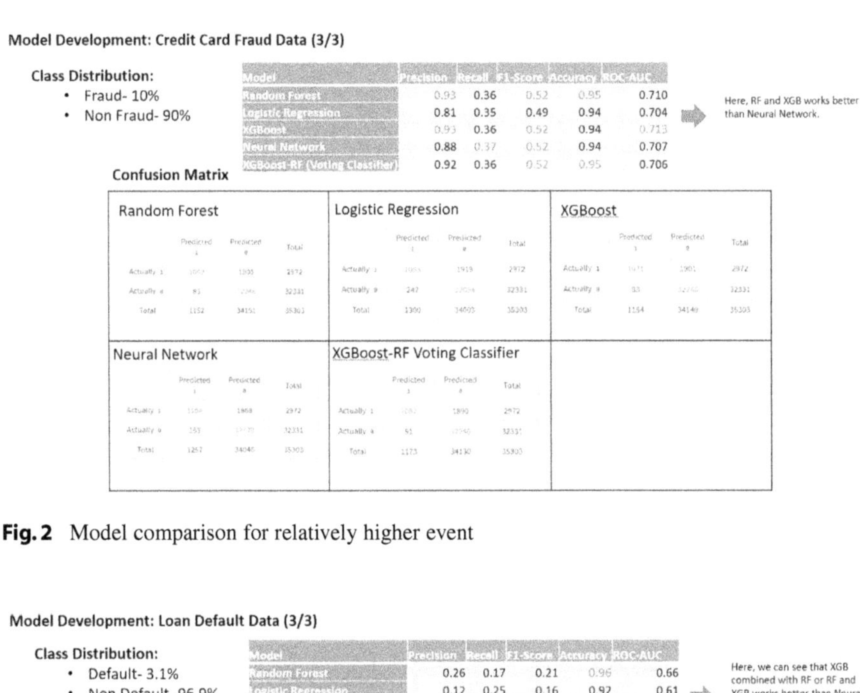

Model	Precision	Recall	F1-Score	Accuracy	ROC-AUC
Random Forest	0.93	0.36	0.52	0.95	0.710
Logistic Regression	0.81	0.35	0.49	0.94	0.704
XGBoost	0.93	0.36	0.52	0.94	0.713
Neural Network	0.88	0.37	0.52	0.94	0.707
XGBoost-RF (Voting Classifier)	0.92	0.36	0.52	0.95	0.706

Fig. 2 Model comparison for relatively higher event

Model Development: Loan Default Data (3/3)

Class Distribution:

- Default- 3.1%
- Non Default- 96.9%

Model	Precision	Recall	F1-Score	Accuracy	ROC-AUC
Random Forest	0.26	0.17	0.21	0.96	0.66
Logistic Regression	0.12	0.25	0.16	0.92	0.61
XGBoost	0.18	0.26	0.21	0.94	0.68
Neural Network	0.11	0.32	0.16	0.90	0.63
XGBoost-RF (Voting Classifier)	0.31	0.18	0.23	0.96	0.67

Here, we can see that XGB combined with RF or RF and XGB works better than Neural Network. Although Recall is high of NN but by scarifying precision

Confusion Matrix

Random Forest

	Predicted 1	Predicted 0	Total
Actually 1	[illegible]	159	228
Actually 0	112	6817	6929
Total	181	7003	7157

Logistic Regression

	Predicted 1	Predicted 0	Total
Actually 1	56	172	228
Actually 0	396	[illegible]	6929
Total	452	6705	7157

XGBoost

	Predicted 1	Predicted 0	Total
Actually 1	[illegible]	169	228
Actually 0	267	[illegible]	6929
Total	326	6831	7157

Neural Network

	Predicted 1	Predicted 0	Total
Actually 1	[illegible]	155	228
Actually 0	594	[illegible]	6929
Total	667	6490	7157

XGBoost-RF Voting Classifier

	Predicted 1	Predicted 0	Total
Actually 1	[illegible]	186	228
Actually 0	93	[illegible]	6929
Total	135	7022	7157

Fig. 3 Model comparison for a higher event rate

To summarize, the choice of models will always be a bit of art and science again. However, a modeler always has a few considerations like level of understanding that the business teams have, degree of linear explainability that the modeler should factor in, sample size, missing values, event rate, and so on.

This appendix can act as a good guide to navigate through the choices before finalizing one of them.

2. Finer aspects of finetuning and finalizing a model

One of the other challenges that the modelers often face is to decide on what is called a good model. Is this good enough. Shall we iterate more or is this good enough. In our consulting context, it is furthermore difficult as we have to deal with the perceptions of the clients. We have faced so many situations where we are presenting the model and its insights after the hard work of almost two months.

Someone from the client team, who has no clue of data sciences will stand up and say this model is not good. When the consultant seek more clarification from the business team around what more could be done, the gentleman say model does not have this variable or that variable. This is one of the most important variables from business perspective.

The concern raised by the business member is a fair one, despite the fact that the timing of this comment is a little late. Ideally, if the business team has an idea of what are the drivers, they should give them to the modeler upfront. So that they can be tested.

Irrespective there are a few genuine concerns that the modeler needs to keep in mind while developing a model:

(a) Are all the factors that should ideally influence prediction of SAR or customer risk or transaction risk are captured. The best way to handle such issue is a brainstorming on all factors and also exploring whether the data underlying those factors is available. It would ensure that realization on relevant factors is not brainstormed at the end, and there is an endless loop during model development process.
(b) We have discussed dimensionality reduction or factor analysis in the model development lifecycle. It is important to use business judgment to prune out any such similar dimension variables and explore more dimensions to make model robust.
(c) The team should test for multicollinearity.
(d) Variables and overall model inference should be similar in training sample and validation sample.
(e) Model should pass the statistical validations including the predictive power and rank ordering of dependent variable when organized in the descending order of the predicted score or probabilities.

The other important question is when to stop iterating. This question is different from number of iterations and stopping criterion that is used for model training.

As a modeler, one can train multiple iterations of the model. But the question would be what is sufficient to stop.

We have partly answered this question in terms of variable selection. Another important aspect is iterations. The variable combination selection will keep happening. However, a modeler will notice that beyond a point, adding more variables is not improving the predictive power. In fact, there will be time when adding a variable is making other variables statistically not significant or the directionality is creating an issue.

That would be the time to question—is the variable introduced makes intuitive sense and whether the added one is more important as compared to existing one which is becoming unstable because of the currently introduced variable.

There are no hard and fast rules, but the basic rule of thumb will still be enough variables that stay stable as a group. Whenever, anything new is added, it does not add to the predictability but result in making this grouping unstable. This would be a good stage to finalize the model. This would be a model that the modeler will be able to defend from variable selection, iterations, and stability perspective.

Appendix 2: List of Dimensions that Are Created as Derived Variables for Model Development and Behavior Analytics

Creation of derived variables
Using the transaction dimensions team created following variables–

1. Deposits
   - Daily deposits
   - Last 7 days deposits
   - Last 15 days deposits
   - Last 30 days deposits
2. Withdrawals
   - Daily withdrawals
   - Last 7 days withdrawals
   - Last 15 days withdrawals
   - Last 30 days withdrawals
3. Deposits as percentage of average of last three months
4. Withdrawals as percentage of average of last three months
5. Deposits as percentage of 90th percentile value of last three months
6. Withdrawals as percentage of 90th percentile value of last three months
7. Percentage change in deposits for 7/15 days
8. Percentage change in withdrawals for 7/15 days
9. Volatility in transaction value for last 7 and 30 days
10. RFM value
11. Velocity wire out/wire in, cash out/cash in.

A. Gupta et al., *Artificial Intelligence Applications in Banking and Financial Services*,
Future of Business and Finance, https://doi.org/10.1007/978-981-99-2571-1

Resources

Threshold fine-tuning of money laundering scenarios through multi-dimensional optimization techniques, Abhishek Gupta, Dwijendra Nath Dwivedi, Ashish Jain, Published 4 March 2021, Journal of Money Laundering Control

A Hybrid Approach for Detecting Suspicious Accounts in Money Laundering Using Data Mining Techniques, C. Suresh, K. T. Reddy, N. Sweta, Published 8 May 2016, International Journal of Information Technology and Computer Science

Identifying Money Laundering Accounts, Chih-Hua Tai, Tai-Jung Kan, Published 1 July 2019, 2019 International Conference on System Science and Engineering (ICSSE)

Counter Terrorism Finance By Detecting Money Laundering Hidden Networks Using Unsupervised Machine Learning Algorithm, A. Shokry, Mohammed Abo Rizka, N. Labib, Published 2020, Computer Science

A Data Mining-Based Solution for Detecting Suspicious Money Laundering Cases in an Investment Bank, N. Le-Khac, S. Markos, M. Kechadi, Published 11 April 2010, 2010s International Conference on Advances in Databases, Knowledge, and Data Applications

A Method to Enhance Money Laundering Detection Using Link Analysis, C. Suresh, K. T. Reddy, Published 2015

A Time–Frequency Based Suspicious Activity Detection for Anti-Money Laundering, U. Ketenci, Tolga Kurt, Selim Önal, Cenk Erbil, Sinan Aktürkoglu, Hande cSerban.Ilhan

Integral representation method based efficient rule optimizing framework for anti-money laundering, Tamás Badics, Dániel Hajtó, Kálmán Tornai, Levente Kiss, István Zoltán Reguly, István Pesti, Péter Sváb, György Cserey, Journal of Money Laundering Control, ISSN: 1368-5201, Article publication date: 18 April 2022

A. Gupta et al., *Artificial Intelligence Applications in Banking and Financial Services*, Future of Business and Finance, https://doi.org/10.1007/978-981-99-2571-1

Data quality issues leading to sub-optimal machine learning for money laundering models, Abhishek Gupta, Dwijendra Nath Dwivedi, Jigar Shah, Ashish Jain, https://www.emerald.com/insight/1368-5201.htm
Efficiency of Money Laundering Countermeasures: Case Studies from European Union Member States Corina-Narcisa (Bodescu) Cotoc 1, Maria Nit,u 2, Mircea ConstantinS, cheau 3,4 and Adeline-Cristina Cozma 1
World Payments Report: Non-cash transaction growth hit hard by COVID, by Alex Rolfe, October 08, 2021
https://www2.deloitte.com/content/dam/Deloitte/tw/Documents/financial-services/tw-the-global-framework-for-fighting-financial-crime-en.pdf
https://www.fincen.gov/resources/law-enforcement/case-examples?field_tags_investigation_target_id=684
https://ai.google/responsibilities/responsible-ai-practices/?category=fairness

Printed in the USA
CPSIA information can be obtained
at www.ICGtesting.com
LVHW020803070923
757371LV00005B/98